PELICAN BOOKS

RELATIVITY FOR THE LAYMAN

James A. Coleman was an instructor in physics and astronomy at Connecticut College (1950–7). He received his B.A. degree in physics from New York University and his M.A. degree in mathematics from Columbia University. From 1947 to 1950 he served as an associate physicist at the Applied Physics Laboratory of Johns Hopkins University, where he did theoretical research on guided missiles and was a consultant to the Research and Development Board on guided missiles. The author also writes a weekly science column for the *Japan Times* and the *Springfield (Mass.) Republican*. His latest books are *Modern Theories of the Universe* (1963) and *Early Theories of the Universe* (1967). James A. Coleman is professor of physics and chairman of the Department of Physics at the American International College, Springfield, Mass.

D0838138

JAMES A. COLEMAN

Relativity for the Layman

A SIMPLIFIED ACCOUNT OF THE HISTORY,
THEORY, AND PROOFS OF
RELATIVITY

REVISED EDITION

Illustrated by the author

PENGUIN BOOKS

Penguin Books Ltd, Harmondsworth, Middlesex, England
Penguin Books Australia Ltd, Ringwood, Victoria, Australia

—

First published by the William-Frederick Press, New York 1954
Published in Pelican Books 1959
Reprinted 1961, 1963, 1964, 1966
Revised edition 1969
Reprinted 1972

—

Copyright © James A. Coleman, 1954, 1969

—

Made and printed in Great Britain
by Richard Clay (The Chaucer Press) Ltd,
Bungay, Suffolk
Set in Monotype Times

This book is fondly dedicated
TO RUTH

Our Purpose

To wonder as we wander
Through this world of mysteries many,
Is to fulfil the purpose
Of our creation and destiny.

And he who does not wonder,
Or the mysteries have not moved,
Should be interred in dampest earth,
For he is dead, and never lived.

J. A. C.

Contents

CONTENTS

List of Figures

Preface

As the title implies, this is a book about relativity, written primarily for those who have had little or no training in mathematics, physics, or astronomy. Technical language is kept to a minimum, and only those mathematical equations which are absolutely necessary for a clear and complete understanding of the theory are included. In short, this book presents the relevant scientific history which preceded relativity, a complete but simple account of the theory itself, and a detailed description of the various proofs of the theory up to the present.

In this age of modern science the education of the individual, regardless of his field, is not complete without a knowledge of the most fundamental branches of science and their historical background. It is particularly important that those of our younger generation – the scientists of tomorrow – be imbued with an understanding of the laws of nature at an early age so that their scientific reasoning and general mental ability will mature side by side. The theory of relativity should be a particularly important part of this process.

The theory of relativity has been considered difficult to understand, but only because its predictions are difficult to believe – not because of any innate incomprehensibility. To introduce such a challenge to the imagination of the young is to stimulate the mind to great accomplishments in the future, particularly where vision is demanded.

The story behind the theory of relativity is a fascinating one which stirs the imagination more than any fiction created by man possibly could do. For here is a story of theory after theory at first appearing successful but disintegrating upon closer scrutiny; of repeated attempts at surmounting insurmountable barriers, only to be met with continual dismal failure. But at last all barriers are surmounted by a superhuman endeavour which up to now has withstood all tests and attacks. This is the story of relativity.

Looking back on the historical developments of the theory, we are particularly fortunate in being able to see the panorama of a hundred years of history flash before our eyes in but a few hours. But more than that, the theory is unique among scientific endeavours because it touches on many phases of the physical world – from the atomic domain to the make-up of the universe itself!

It is my belief that readers will derive as much enjoyment and inspiration from this book as I did in writing it.

17 March 1954   J. A. C.

Preface to Revised Edition

IN the fifteen years since this little book was originally written many developments have occurred which relate to relativity. In most cases these have been the application of later scientific discoveries and new techniques to the original predictions of the theory. In the case of the time, or clock, paradox the long-smouldering disagreement regarding its explanation finally erupted in 1957–59 in a voluminous outpouring of papers on the topic. I have taken advantage of the clarification of this topic which has resulted and have completely rewritten the section dealing with it to conform to the present views of the majority of those theoretical physicists who are deemed to be most knowledgeable about relativity.

In all cases the more recent tests of the theory of relativity have, without exception, further verified the basic predictions of the theory. The fundamental theory and predictions as originally announced by Einstein thus remain unchanged.

Some of the changes made in the text represent an attempt at further clarification of points raised by many readers who have written me. While every attempt has been made to make this revision as clear as possible, it is not claimed that the result is the penultimate introduction to relativity. So readers are invited to continue to forward any suggestions for further improvement or any questions which still remain after a careful reading of the text.

American International College   J. A. C.
Springfield, Massachusetts 01109
September 1968

1 · The Velocity of Light

Mersenne's Measurement of the Speed of Sound

SINCE the theory of relativity really had its beginning in the peculiar behaviour of light waves, we will start our study with the history of one of the most important properties of a light wave, namely, its *velocity*. But a word about the velocity of *sound* first, since this was measured before the velocity of light. The ancients seemed to have been aware, and correctly so, that when something made a noise the sound travelled from the thing which made the noise to the ear of the listener. This conclusion stemmed in part from observing that the farther away one was from a flash of lightning the longer it took for the noise of the ensuing thunder to be heard. However, the velocity with which the sound wave travelled was not measured until the Middle Ages.

One of the earliest measurements of the velocity of sound was made by Mersenne (1588–1648), a Frenchman. He set up a cannon several miles distant from him and had his assistant fire it. From his observation post he measured the length of time it took the noise of the blast to reach him after he saw the flash, by counting the swings of a pendulum during this time. (A swinging pendulum was used to measure time intervals, since stop watches were not yet in use.) Since he knew how long it took the pendulum to make one swing, he computed the total time it took the sound to travel from the gun to him and divided this into the distance between him and the gun (which he had previously measured); thus he derived the speed of sound. His result was approximately 700 miles an hour. More exact methods now show it to be about 750 miles an hour. This was considered a very high

13

speed in Mersenne's day when we remember that a fast race-horse can run at about 40 miles an hour. Today aeroplanes fly considerably faster than this, with some flying even faster than the speed of sound, not to mention guided missiles which have velocities several times that of the speed of sound.

Galileo's Attempts to Measure the Velocity of Light

Consider what happens when we enter a dark room and turn on the electric-light switch. As far as we are able to tell, the light coming from the light bulb hits our eyes instantly. But if we investigate what happens, we must agree that the source of the light is the light bulb itself, i.e., the light that floods the room first must come from the bulb. Hence, we are forced to conclude that the light travels from the light bulb to our eyes to give us the sensation of light. But our senses seem to tell us that we see the light the very instant we throw the switch. We now know that the speed of light is so great as to make it appear to travel instantaneously from one place to another.

The battle over whether the velocity of light was infinite or finite raged with full fury during the Middle Ages, with no less eminent a scientist than Descartes (1596–1650) claiming it to be infinite, while Galileo (1564–1632), another great scientist of the day, claimed the velocity to be finite.

In an effort to prove that he was right, Galileo attempted to measure the velocity of light. One dark night he stationed a co-worker, equipped with a lighted lantern covered with a pail, on a hilltop 3 miles away from him. Galileo also had a lantern covered with a pail. When both were ready, Galileo lifted his pail, thus permitting the light rays from his lantern to travel towards his assistant with the velocity of light. The assistant, upon seeing the light, also lifted his pail, and the light rays from his lantern in turn travelled back to Galileo – again with the velocity of light. Galileo measured the total time from when he first lifted his pail to when he received the light rays from his helper's pail and, having measured the distance between the two positions as accurately as possible beforehand, he then computed the velocity of light.

Each time Galileo did the experiment he obtained a different value for the velocity, and so the results of the experiment were inconclusive. We now know the reason the experiment failed: the time it took Galileo and his assistant to notice each other's lanterns and then act, i.e., their reaction time, was so long in comparison with the travel time of light that the light rays from their lanterns could travel completely around the earth fourteen times if we assume their reaction time was one second each. We see that although the method Galileo used appeared sound at the time, it was as futile as for a snail to try to catch a fly.

Roemer's Astronomical Method

What was necessary was either to time the passage of a light beam over a large distance – greater than the circumference of the earth – or to use a shorter distance, provided that an accurate clock was available. Quite by accident, an astronomical method presented itself not too many years after Galileo's futile attempt and, ironically, one of Galileo's early discoveries in astronomy made the opportunity possible.

Galileo had built a telescope, one of the first made, with which in 1610 he discovered the four most prominent moons of the planet Jupiter. (Jupiter has twelve known moons.) Like our own moon, each of them travels in an orbit around the planet, each in its own uniform amount of time called its *period*.

In 1675 Olaf Roemer, a Danish astronomer, measured the periods of these four moons of Jupiter, but he obtained different results when he measured them again several months later. To understand what happened, see *Figure 1*. The left side shows the earth at two different positions in its orbit around the sun, while the right side shows the planet Jupiter and its first moon, which has just been eclipsed by the planet. (Although Jupiter also revolves around the sun, its period is so much larger than that of the earth that we can neglect its motion here.) Roemer measured the period of this first moon by timing how long it took for the moon to go from

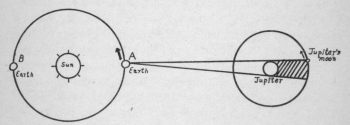

Figure 1. Roemer's Determination of the Velocity of Light

the position shown, around in front of Jupiter, and back again to this same position. He found the period to be approximately 42½ hours when the earth was at position *A*. Now, since it is known that the period is constant, we can predict that succeeding eclipses will occur every 42½ hours. But as the earth travelled in its orbit away from the planet Jupiter in going from *A* to *B*, Roemer found that the eclipses were occurring later and later, until in six months, when the earth was in position *B*, the eclipse occurred 1,000 seconds too late!

The only logical conclusion Roemer could draw was that this additional time represented the time it took the light from Jupiter's moon to travel the extra distance across the diameter of the earth's orbit. At that time the diameter of the earth's orbit was believed to be about 172,000,000 miles, instead of the correct value of about 186,000,000 miles, so that Roemer's data produced too low a value for the velocity. However, Roemer's method is remembered historically as the first successful determinaton of the velocity of light.

Bradley's Telescope Method

The next determination of the velocity of light was made by a different astronomical method in 1727 by James Bradley, an Englishman. To demonstrate his method, we first consider a simple analogy with which most of us are familiar. Suppose you are on a train that is about to start. It is raining outside. You will notice that the rain runs down the window-pane more or less in a straight path from top to bottom. This, of course, is as it should be. But now the train begins to move. You notice that the rain streaks no longer run straight down the glass but fall at an angle. The streaks start at the top of the glass and run towards the rear of the pane, as in *Figure 2(a)*. And the faster the train moves, the more horizontal the streaks become. The amount of tilt of the streaks, then, is related to the speed of the train.

The explanation of what has happened is easy to understand. Going back to the vertical streaks the rain makes when the train isn't moving, we would notice, if we were to make some measurements, that all the rain-drops run down the window-pane with approximately the same velocity. As a result, it takes the same amount of time for each drop to run down the pane. Now, when the train is moving, the drops still fall in the vertical direction with the *same* velocity because the forward motion of the train does not affect the speed with which the drops are falling towards the earth. But during the time the drops are falling from the top of the window to the bottom, the train moves forward. Hence it appears as if the drops are running backwards to you, since you are on the moving train.

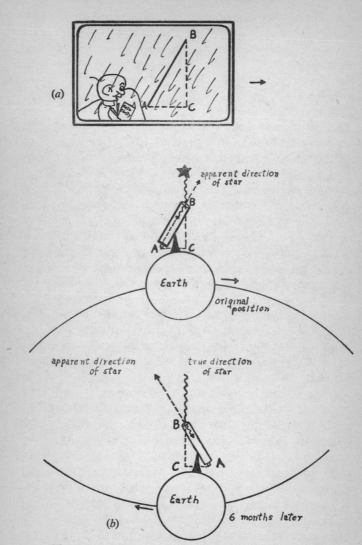

Figure 2. Bradley's Determination of the Velocity of Light

It should now be clear why the streaks become more horizontal the faster the train moves. During the same time it takes a drop to fall, the train moves an even greater distance forward and the drops move a greater distance backward on the window-pane.

Perhaps it has now occurred to you that it should be possible somehow to compute the speed with which the drops fall towards the earth if we know the velocity of the train and use a ruler to measure the lengths of the sides of a right triangle, such as *ABC* in *Figure 2(a)*. It can be easily shown by elementary trigonometry that the velocity of fall of the drops is the product of the train's velocity by the ratio of the sides *BC* to *AC*. The method that Bradley used to determine the velocity of light is closely analogous to this.

Bradley's method is illustrated in *Figure 2(b)*. We will assume we have a telescope and want to look at a distant star. Here, the light 'falling' to the earth from the star represents the falling rain-drops; the motion of the earth in its orbit represents the moving train; and the telescope through which the starlight falls represents the window-pane. (The sun is purposely omitted.) If we want to see the star through the telescope, we must aim it so that the light coming from the star 'falls' down through the telescope barrel and hits our eye at the viewing end.

If the earth were stationary in space and did not move in an orbit, we would point the telescope straight up at the star and its light rays would 'fall' straight down through the centre of the telescope. This corresponds to the previous example of the train standing in the station. But we know that the earth does move in its orbit around the sun with a velocity of about nineteen miles a second. Hence, the actual situation is as shown and the telescope must be tilted so that when the light waves from the star enter the telescope at *B*, they will travel down the centre of the barrel and meet the eye at *A*, in the same way that the paths of the rain-drops were tilted on the window-pane. While the light waves are going from *B* to *C*, the observer (and telescope) will move from *A* to *C*, since he is on the moving earth.

But here the similarity to the falling rain-drops ends. We cannot compute the velocity of the incoming light waves as we did that of the rain-drops. Although we knew that the rain-drops came from directly overhead, we do not *know* that the star is directly overhead. Since we see it in the aimed telescope, it looks as if the star is in the direction the telescope is pointing, i.e., along the line *AB* extended. This is the situation that existed up to Bradley's time. No one ever dreamed that a star could be in any other direction but that in which the telescope pointed.

But Bradley noticed that six months later the same star appeared to be in a different direction in the sky. He called the phenomenon *aberration* and explained it as we have done, asserting that the velocity of light must be finite as a consequence. He used the star Gamma Draconis and found that the variation in its direction over the appropriate six-month period was about forty seconds of arc, or about one ten-thousandth of a right angle. This told him that the angle of tilt was about twenty seconds of arc from the vertical. Then, knowing the angle of tilt, he constructed the right triangle *ABC* and computed the velocity of light in the same way we did for the rain-drops. Here, similarly, the velocity of light equals the product of the earth's orbital velocity and the ratio of the sides *BC* to *AC*.

Bradley's result wasn't too accurate, but his method was important because it lent greater credence to the fast-growing belief that the velocity of light certainly was finite even though it was as great as 186,000 miles a second. But Bradley's experiment was also important because it was one of the experiments which led directly to the theory of relativity, as we shall see.

Fizeau's Terrestrial Method

The first determination of the velocity of light which didn't use astronomical methods was made by Fizeau in 1849. He supplied what was lacking in Galileo's attempt, i.e., the accurate measurement of the small interval of time it took a light beam to cover a relatively short distance on earth.

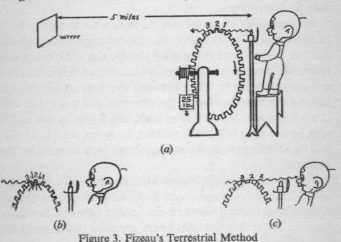

Figure 3. Fizeau's Terrestrial Method

His method is illustrated schematically in *Figure 3*, showing a cog-wheel being turned by a weight-and-pulley arrangement (there being no electric motors in Fizeau's day). The candle is the source of light waves whose velocity is to be measured while they run a course from the candle to a mirror five miles away and back again.

If we first assume the wheel is not moving, then light from the candle will go between cogs 1 and 2, will travel the ten-mile round-trip from the cog-wheel to the mirror and back, then go through the same opening between the cogs, and will be seen by the eye behind the candle. But now suppose the wheel is moving as shown in (*a*). Then the beam

21

from the candle will be chopped up by the cogs as they pass before the candle, in the same way as a meat slicer slices bologna. The result will be a series of pieces of light or individual beams sent towards the mirror, their exact length depending on how fast the wheel is rotating; the faster it rotates, the shorter the beams. (Only the beam that originally starts out between cogs 1 and 2 is shown in the drawings for simplicity.)

Now consider what happens when this beam arrives back at the wheel after travelling the ten-mile round-trip to the mirror. If the wheel is rotating slowly, the light beam will arrive while cog 2 is in front of the candle, as in (b), and will not get through the space between the cogs to hit the eye beyond the candle. Hence, the observer will not see it. But if the wheel is rotating fast enough, cog 2 will be out of the way by the time the beam returns, as in (c), and the beam will be able to pass between cogs 2 and 3 to be seen by the observer.

This is exactly what Fizeau did. Starting with his wheel at rest, he gradually increased its speed until he saw the reflected light coming through the spaces between the cogs. He then knew that the light beam travelled ten miles in the time interval it took for one space between two cogs to be replaced by the next one. He computed this short time interval by measuring the speed of rotation of his wheel and knowing the number of cogs in the wheel. Then he divided the total distance the beam had travelled by the time interval and obtained 194,600 miles a second for the speed of light – a result about 5 per cent too high, but quite accurate considering the limitations of his equipment.

Michelson's Precise Measurement

The most widely known measurement of the velocity of light is that performed by Michelson (1852–1931) in 1926. This experiment was not only heralded as an accurate determination but also stands as a monument to experimental technique. Some of the difficulties encountered were almost beyond solving, but Michelson was equal to the task, having been the first American to win the Nobel Prize in physics (1907). Michelson perfected the revolving mirror method which had been used by Foucault in 1850. This is somewhat similar to Fizeau's cog-wheel but uses a rotating, many-sided mirror to chop up the original light wave into individual beams which, like Fizeau's, are sent to a distant mirror (here 22 miles away) and back again (see *Figure 4*). The many-sided mirror is represented by one with six sides and can be rotated at any desired speed with an electric motor.

We first consider what would happen if the mirror was not rotating and set at position (*a*). Light would leave the light source (represented here by a light bulb) and hit side 1, from which it would be reflected to the distant mirror. Upon reflection there, it would return via the same general path, hit side 1 again, and would be seen by the eye near the lamp as it returned towards the general direction of the lamp.

But now suppose, as happens in the actual experiment, that the mirror is rotated at the instant the beam leaves side 1 on its journey towards the distant mirror, as in (*a*). If the speed of rotation is not fast enough for side 2 to be in the same position as side 1 originally was by the time the beam returns, then the beam will not be reflected to the observer's eye but in some other direction, as in (*b*). But if the speed of rotation is such that side 2 is in the same position as side 1 originally was by the time the reflected beam returns, then the beam will hit the observer's eye, as in (*c*).

When this is the case, then during the same time interval it takes the beam to make the round-trip to the distant mirror and back, the rotating mirror rotates by a sixth of a

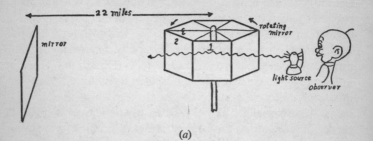

(a)

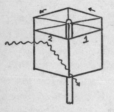

(b)

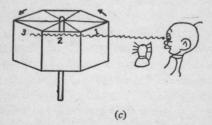

(c)

Figure 4. Michelson's Precise Determination

24

revolution. Further, since the speed of rotation of the mirror is known, the time for one revolution is known, and one-sixth of this is the time it takes for the light beam to make its round-trip. Dividing the round-trip distance by this time interval gives the velocity of light.

In Michelson's actual experiment various rotating mirrors of 8, 12, and 16 sides were used. This apparatus was set up on Mount Wilson in California. The mirror system comprising the distant mirror was set up on Mount San Antonio, approximately 22 miles away. Since the accuracy of the results depended greatly on the precision with which this distance was known, the U.S. Coastal and Geodetic Survey measured it exclusively for Michelson's experiment, with an error of less than 2 inches. This alone was almost a superhuman accomplishment. Due to the painstaking care which Michelson exercised in all phases of the experiment, the results are considered accurate to within a very small fraction of 1 per cent. As a result of this and Michelson's subsequent experiments, we now know that the velocity of light is approximately 186,000 miles a second – and this is the value we will use throughout this book.

Other Properties of Light Waves

One hundred and eighty-six thousand miles a second may seem like too high a speed to imagine, but it can be put into ideas with which we are familiar. For example, light will go completely around the earth in about a seventh of a second. And it takes about eight minutes for sunlight to travel the 93 million miles from the sun to the earth. Thus, in the morning when you see the sun come up, it has actually risen eight minutes earlier, and in this respect no one on earth ever sees the sun rise when it does!

Since it takes eight minutes for the sunlight to travel from the sun to the earth, we can say that the sun is eight minutes away by light in the same way that we say Stamford is forty minutes from New York City by rail. Astronomers use such a time-table for listing the vast distances of the various stars from us, since the distances expressed in miles are too large to list easily. The nearest star to us beyond the sun, for example, is Alpha Centauri, which is about four light years away. This means, of course, that it would take us four years to get there if we travelled by light, i.e., with the speed of light. Since one light year is about 6,000,000,000,000 miles, Alpha Centauri is about 24,000,000,000,000 miles away, which is indeed a long way for a next-door neighbour! Using this same nomenclature, we would say that Stamford is forty rail minutes from New York City.

In discussing the experiments for measuring the velocity of light, we made no distinction between the velocity of light in vacuum (which is what most of outer space is) and other materials – air, for example. It is natural to think that light would travel faster through vacuum than through air, since we would expect that light would be less impeded in vacuum with nothing there to slow it down. Such is the case. However, light travels only slightly slower in air, so that for all practical purposes it can be assumed that its speed is the same in air as it is in vacuum. Not so for denser materials like water, for instance. Here, the velocity of light is only three-fourths what it is in vacuum (or air), and in glass it is only two-thirds.

We have talked about light waves without mentioning other waves or rays, such as radio waves, infra-red rays, etc. It so happens that light waves, radio waves, ultra-violet rays, infra-red rays, etc., are all part of a general group called *electromagnetic waves*, and all members of the group travel with the velocity of light. We shall deal mainly with light waves in this book, since these are the only ones that are visible.

2 · The Great Dilemma

The Stationary Ether Postulated

WITH proof afforded by many excellent experiments that light travels with a finite velocity of about 186,000 miles a second, scientists next turned their attention to the consideration of the medium which carried or propagated the light waves. This prelude to the theory of relativity took place during the period from 1800, by which time the finite velocity of light was firmly established, to 1905, when Einstein introduced his Special Theory of relativity.

It was known that sound waves are propagated by setting the air (or other material through which they travel) into vibration. This vibration, or wave, is in this way pushed forward. It was further found that sound waves could not travel through a vacuum – some material substance was necessary for their propagation. Then, too, water waves needed water in which to travel; a water wave without water to carry it could not exist. Such was the reasoning used. As a result, it was believed that light waves also had to have a carrier or tangible substance in which they could push themselves forward.

However, it also was known that out in the vast reaches of space between the planets and stars there was no air or other medium – most of space was vacuum. But, nevertheless, no one could doubt that light did travel the 93,000,000 miles from the sun to us through this vacuum. Not wanting to believe that light travelled through nothing – which carried the implication that no medium of any kind was necessary for its propagation – scientists created a special word for the hypothetical carrier of light waves. They called it the *lumeniferous ether*, or just plain ether. The ether, then, was the material that existed everywhere that light waves travelled, and these light waves moved through it at a velocity of 186,000 miles a second. The ether filled the vast emptiness of the universe and was present in all substances in greater or lesser degree. The idea of the existence of the ether seemed so logical that it quickly gained widespread acceptance as one of the materials in the universe. And some scientists even went so far as to determine its density theoretically!

Further Confirmation of the Ether

Additional confirmation for the existence of the ether came unexpectedly from the realm of electric and magnetic phenomena (or, more strictly, electromagnetic theory). In 1864 Maxwell published the results of a mathematical investigation he had undertaken on electrical vibrations. He showed that certain electrical vibrations would cause electrical waves to be formed which would travel outward through space. Furthermore, he calculated the velocity with which these waves would propagate and found it to be 186,000 miles a second – the same speed earlier scientists had determined for the velocity of light! Maxwell then correctly concluded that light waves were nothing more than a particular type of his electric waves or, as we call them today, electromagnetic waves. And in 1887, Maxwell's prediction of the existence of the electromagnetic waves was verified when Herz succeeded in generating them in the laboratory.

With Maxwell's discovery that light waves were electromagnetic in nature, the necessity for the existence of the ether was further strengthened, since it was believed that electric and magnetic fields must have a substance in which to reside, it being inconceivable that they existed in a vacuum. Electromagnetic waves as a group, then, certainly had to have a medium to carry them, and the ether was the only logical medium.

After the idea of the existence of the ether had become firmly entrenched, effort was directed to the detection of the ether; and it was here that science met its nemesis, as we shall see.

If the ether existed, then, since it permeated all space, it was reasoned that the ether was the one thing which remained fixed in the universe and did not move. It was known that the earth and other planets were not stationary with respect to the sun; in particular, the earth was known to revolve about the sun with a velocity of about nineteen miles a second. It was not known just how stationary the sun was with respect to the other stars, but it was believed that the ether alone remained motionless in the background of the moving heavenly bodies in much the same way water in a goldfish bowl remains motionless while the fish swim about in it.

Scientists asked themselves: if all the heavenly bodies are moving with respect to one another, how can we tell if they are moving about in the ether, which itself remains motionless? If you find yourself on a ship out at sea and want to know whether or not you are moving, you look to see if water is moving by the ship. It is easy to decide – you can see the bow wave, or you can put your hand into the water, and if the water flows around it you conclude that you are moving through the water. This is precisely how scientists proceeded to detect the ether – by attempting to discover the *ether drift* or *ether wind*, as it was called. If the ether wind could be found, it would be proof not only that the earth moved through the ether but, what was more important, that the ether existed as it was believed. Unfortunately, the ether wind could not be detected merely by sticking one's hand out into space and feeling for it.

An Expected Ether Effect

There were several effects that should exist if the ether wind existed, and these were looked for eagerly. We will now discuss one of these effects, duplicating the reasoning used at the time it was looked for. Assume we have a telescope set up on the earth (*Figure 5*). We focus it on a star which is in

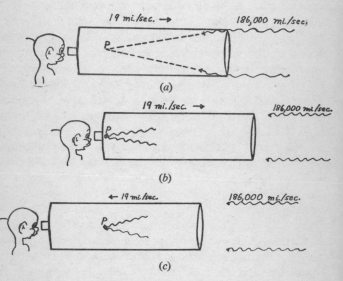

Figure 5. Expected Change in Focus of a Telescope Due to Motion through the Ether

the direction the earth is travelling in its orbit. The light from the star is travelling at 186,000 miles a second through the assumed stationary ether between the star and the earth. Two of the light beams from the star have just entered the telescope in (*a*). These beams have been bent by the telescope lens so that they will come to a focus at point *P*, which is a point in the space within the telescope. Now since the telescope and observer are moving to the right with a

30

velocity of 19 miles a second, the observer's eye will arrive at point P at the same time the light beams do in (*b*), and the observer will see the star in focus.

But now suppose the astronomer looks at the same star six months later and does not change the focus. The situation will be entirely different, since the earth will be on the other side of its orbit. Whereas before it was travelling towards the star, or to the right with respect to the ether, with a velocity of 19 miles a second, it will now be travelling away from it, or to the left with respect to the ether, with the same velocity. What was expected to happen is shown in (*c*). Since the telescope and observer are now running away from the incoming light wave, the observer's eye will no longer be at point P when the light beams arrive there, and as a consequence the observer will now see the star out of focus. If this reasoning were correct, then a telescope which was originally in focus on a distant star would be out of focus six months later. This effect was looked for but was never observed.

Fresnel's Ether Drag

A possible explanation for the failure to detect this effect was contained in a theory advanced by Fresnel in 1818. He believed that the ether was thicker in material bodies than it was in vacuum or outer space and that, as a result, when a transparent object such as a telescope lens moved through the ether, it dragged some of the ether along with it in much the same way a moving ship drags some water behind it. On the basis of this assumption, Fresnel computed the amount of the ether drag as a certain fraction of the velocity of the moving object, in this case the telescope lens. This fraction came to be known as the *Fresnel drag coefficient*.

The net effect would be that whether the telescope was travelling towards the incoming light waves or away from them, the ether would be dragged along with the telescope; it would be impossible to detect the effect, since to do so it would be necessary for the ether to stay put while the telescope moved through it. This is similar to hanging a fish from a pole and tying it to a running dog so that it dangles in front of him: the dog never catches the fish because it moves whenever he moves.

Since the Fresnel drag coefficient was only theoretical and had no direct experimental proof to support it (except the indirect evidence presented by the inability to detect the effect), an experiment was needed which would measure the velocity of light in a fairly dense material which was itself moving. This was done by Fizeau in 1859. He used moving water and measured the velocity of a light beam travelling through the water in the same direction as water movement, and then again in the direction opposite to the water movement. He found that the velocity of light in the water was affected by the fact that the water was moving, i.e., the result was as if the water dragged the ether along with it by the same amount given by the Fresnel drag coefficient.

The reader should not believe that as a consequence of all this it was proved that the ether really existed and that it was dragged along with a moving object so that it escaped detection. The Fresnel ether-drag theory was a possible explanation *if, and only if*, the ether existed and behaved as outlined.

The Michelson–Morley Experiment

When it was found that the ether's existence could not be detected by changing the focus of a telescope over a six-month period, as well as by other similar effects, the ether's existence was by no means doubted. What was needed, it was said, was a much more sensitive experiment – one which would definitely show up the ether's presence. Such an experiment was devised and carried out by Michelson and Morley in 1881.

Before going into the details of the Michelson–Morley experiment to detect the ether wind, we will first consider an analogy whose fundamental reasoning is that used in their experiment. We will assume we are going to race two

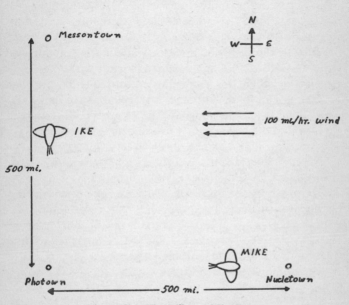

Figure 6(*a*). The Mike and Ike Race

identical planes, Mike and Ike, against each other, starting both at the same time from the same place, Photown, as shown in *Figure 6(a)*. We will have Mike go due east to Nucletown and back while Ike goes due north to Messontown and back. We will assume that both Nucletown and Messontown are exactly 500 miles from Photown. Now, if the top speed of both Mike and Ike is 1,000 miles an hour and if there is no wind at the time of the race, the reader would expect the race to end in a dead heat in an hour – which, of course, it would.

But suppose that all during the race there was a 100-mile-an-hour east wind blowing: the race would not end in a tie because Ike would win. The reason is that while Mike is travelling east towards Nucletown, the 100-mile-an-hour wind is permitting him to make only 900 miles an hour over the ground. (The 1,000-mile-an-hour maximum speed of the planes is with respect to still air.) Returning, however, Mike is aided by the east wind and does 1,100 miles an hour over the ground. But since he travelled a longer time at the slower speed while going, his average speed for the trip is less than 1,000 miles an hour. Although it is true that Ike had a side wind of 100 miles an hour both going and coming and had to turn into the wind slightly to compensate for it, the wind did not slow him down as much as it did Mike. Ike also averages slightly under 1,000 miles an hour, but still higher than Mike.

This reasoning can be verified algebraically if the reader is so inclined. It turns out that for the particular values used it will take Ike eighteen seconds over an hour to go to Messontown and back, but it takes Mike thirty-six seconds over an hour to go to Nucletown and back. Hence, Mike will return eighteen seconds later than Ike, and *Ike always wins*.

So far we have not connected the race between Mike and Ike with the Michelson–Morley experiment. The connexion is this: If the velocity and direction of the wind were unknown on the day of the race, it could be determined by the finishing position of Mike and Ike on their return to Photown. If they both returned simultaneously at the end of an hour, you would conclude that there was no wind. But if Ike came back in an hour and eighteen seconds and Mike in an hour and thirty-six seconds, it would indicate there was a 100-mile-an-hour wind in the east–west direction. (It could not be determined whether it was an east wind or a west wind, but this is unimportant here.) And if Mike and Ike were to interchange courses, then Mike would be back in an hour and eighteen seconds while it would take an hour and thirty-six seconds for Ike.

Thus, one way to detect the wind would be to have Mike and Ike race, then interchange courses and race again. If there is a shift in their finishing positions, a wind is present, and the greater the shift the stronger the wind. This, in

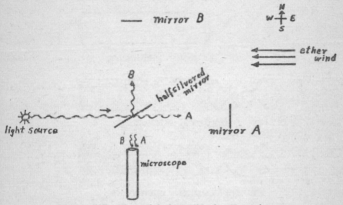

Figure 6(*b*). The Michelson–Morley Experiment

effect, is what Michelson and Morley did. They raced two light waves at right angles to each other; then interchanged their courses, raced them again, and looked for a shift in their finishing positions. Such a shift would conclusively prove the existence of the ether wind.

The physical apparatus used by Michelson and Morley is shown in *Figure 6(b)*. If the earth is moving to the right with respect to the ether, we would experience an ether wind in the direction indicated. A light wave from the light source strikes the half-silvered mirror which splits the light wave into two equally intense waves, A and B. The A wave goes through the half-silvered mirror and on to mirror A, while the B wave is reflected at the half-silvered mirror to mirror B. These two individual light waves correspond to Mike and Ike. The A wave will be reflected by mirror A, will return to the half-silvered mirror, and half of it is reflected to the microscope where the observer views it. (The other half of the A wave goes back to the source, but this is unimportant to the experiment.) Similarly, the B wave is reflected at mirror B to the half-silvered mirror, and then half of it

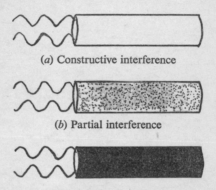

(a) Constructive interference

(b) Partial interference

(c) Destructive interference

Figure 7. Constructive and Destructive Interference

also goes to the observer's microscope. The observer then sees both waves in his microscope and notes their 'finishing position'.

He then interchanges the courses of the A and B waves by rotating everything through 90°, either clockwise or counterclockwise. The A wave will now travel in the north–south direction while the B wave travels in the east–west direction. The observer again notes their finishing position and compares it with the previous 'race' to note if it has shifted.

To determine whether or not the finishing position has shifted, the observer uses a phenomenon of wave motion called *interference*. To illustrate, if the two waves enter the microscope (see *Figure 7*) so that the hills and valleys of each are lined up, or are in phase, as in (*a*), the waves tend to reinforce each other, i.e. the viewer sees the resultant light wave brighter than either individual wave. The result is called *constructive interference*. If one wave is slightly behind or ahead of the other, as in (*b*), they do not reinforce each other quite so much, and the viewer sees the resultant light dimmer than it was before. But if the waves are sufficiently out of phase so that a valley of one is lined up with a hill of the other, the hills and valleys interfere, cancelling each other out so that darkness results, as in (*c*). This is referred to as *destructive interference*.

Since the device used by Michelson and Morley employs interference phenomena, it is called an *interferometer*. Now when the observer rotates the interferometer through 90° if there is an ether wind present, it should cause the relative finishing positions of the waves to change, i.e., one wave should shift with respect to the other. And this shift will cause the light in the microscope to change, becoming brighter or dimmer as the case may be.

When Michelson and Morley performed the experiment they did not detect any change whatsoever in the light intensity in the microscope upon rotation, which meant that they did not detect any ether wind. They repeated the experiment at different times of the day and during various times of the year, but the results were always the same – they did not detect an ether wind. The Michelson–Morley experiment has been repeated a number of times with increasing accuracy. The most recent, and by far the most accurate, search for the ether was reported in 1960 by Professor Charles H. Townes of Columbia University, the inventor of the *maser*, with the help of physicist John Cedarholm. The main feature of the maser of importance here is that when certain molecules are excited electrically they emit microwaves of very stable and accurately known frequency. In fact, two ammonia masers can be made so stable in frequency that they would not vary by as much as one second for at least 200,000 years!

Two ammonia masers were set up so that the beams they emitted were oppositely directed and in the east–west direction. After about one minute the two masers were interchanged. This was done a number of times throughout a 24-hour period so that any changes due to the earth's rotation could be noted. The experiment was repeated for a number of days at intervals of three months throughout a year.

The basic theory was that any effect due to the motion of the earth through the ether would be indicated by a change in the frequency of one or both masers. The frequencies of the masers were thus recorded continuously so that they could be compared throughout the experiments. Computation showed that the effect of an ether, if it existed, would be to produce a difference in frequency between the two masers of about 20 cycles per second. But no such frequency change was noted. In fact, the experiment was so precise that if any ether effects were present they would have been detected even if the earth's orbital velocity were only one one-thousandth of what it actually is.

Modern science has thus overwhelmingly verified the conclusion of Michelson and Morley, and it is now universally accepted that the ether cannot be detected.

Possible Explanations for Michelson and Morley's Results

The failure to detect the ether could be explained, of course, if the ether did not exist; but the necessity of the ether's existence was too firmly entrenched to be discarded. Instead, four reasons were advanced as possible explanations of the inability of scientists to detect the ether. The easiest explanation was that the earth was fixed in the ether and that everything else in the universe moved with respect to the earth and the ether. Then we on earth would not experience an ether wind, thus making the detection of the ether impossible. Such an idea was not considered seriously, since it would mean in effect that our earth occupied the omnipotent position in the universe, with all the other heavenly bodies paying homage by moving around it. The fact that the earth was only one of several planets revolving around the sun was enough to dispel any notion that, as a planet, it occupied any kind of godly post.

It was also thought possible that the earth *dragged* the ether next to it along with it. This, too, would make an ether wind impossible. But there were two insurmountable objections to this explanation: if the ether were dragged along with the earth, then light waves coming into the earth's vicinity would also be dragged along, since they travel in the ether. But if so, we would always see light waves from a distant star coming from the same direction, and we would not observe the aberration phenomenon discovered by Bradley.

It will be recalled that the apparent direction of a star changes over a six-month period, since the earth has a velocity of 19 miles a second in its orbit with respect to the incoming light from the star. If the ether were moving along with the earth, the light from the star also would be swept along with it and the star always would appear to be in the same direction. But since we know that the star's direction does change, i.e., that aberration does exist, we know that the ether cannot be dragged along with the earth.

The other objection to this possibility is concerned with the Fresnel drag coefficient. As we mentioned earlier in this chapter, it was found that some materials did act as if they dragged the ether along with them, but this was only a partial drag, i.e., the ether appeared to come along with only a fraction of the velocity of the moving object. Here, however, it would be required that the ether be dragged along at the full velocity of the moving earth. Furthermore, it was not known whether an object as large as our earth would conform to the Fresnel drag coefficient, since Fizeau's verification of the drag effect was made on a laboratory scale only.

The third possible explanation for the inability of the Michelson–Morley experiment to detect the ether assumed that the velocity of light was always constant with respect to the source which emitted it. This would mean that light always travelled at 186,000 miles a second with respect to the interferometer, regardless of how fast or slow it was moving with the earth through the ether. As a result, the velocity of light would vary with respect to the ether. The ether would not be detected because both light beams would always have the same velocity with respect to the interferometer, and any race between them would always end in a dead heat. Going back to the analogy of Mike and Ike, it would be as if they both always had the same velocity with respect to the ground, regardless of whether or not a wind was blowing.

The main objection to this third explanation was that it required the velocity of light to vary with respect to the ether. This was contrary to the generally accepted notion of wave motion that the velocity of the wave must be constant in the material which carried the wave. Sound waves travelling through the air were used as the classic example. It was well established that the velocity of sound waves was independent of whether or not the source of the sound was moving. It was thus difficult for anyone really to believe that the velocity of light through the ether was influenced by the velocity of the source. Indeed, the ether had originally been postulated as the carrier of the waves, and one of the reasons for so doing was to create a medium with respect to which light would always have a constant velocity.

There were also various astronomical observations which indicated that the velocity of light was independent of the velocity of the source. One of these was in connexion with double stars. Double stars are two stars which are approximately the same size and are relatively close together. They rotate about each other with a fairly high velocity in somewhat the same way as would the ends of a dumb-bell if the dumb-bell were thrown into the air so as to rotate end over end. Now, some of these double stars rotate so that we are looking edgewise at the plane of rotation, i.e., we see one star coming towards us while the other is going away, and vice versa. If we assume that the velocity of the light waves leaving the star is increased or decreased by the velocity with which the star is approaching or receding from us, then the star approaching us would appear to be rotating much faster than the receding one. And when their positions are reversed, the situation would also reverse. The overall effect would be as if the stars were alternately speeding up and slowing down in their rotation about each other. Actual observation shows that this is not the case, however, and that the stars actually rotate about each other with uniform velocities. We conclude that it is entirely unlikely that the velocity of light through the ether is influenced by the velocity of its source, or that it is constant with respect to the source.

The explanation which had the most appeal in accounting for the negative result of the Michelson–Morley experiment was one that was literally dreamed up for the purpose. It is the so-called *Fitzgerald–Lorentz contraction.* In 1893 Fitzgerald suggested that all objects contracted in the direction of their motion through the ether. He reasoned that if ordinary objects flattened out upon impact with other objects – a rubber ball hitting a wall or a ripe tomato dropped on the floor, for example – then why would it not be possible for objects that move through the ether to have the force of the ether push them in, or contract them? This would adequately explain the results of the Michelson–Morley experiment. The arm of the interferometer moving against the ether would be shortened so that, even though the light wave travelling in that particular arm might be slowed down by the ether wind, this would be compensated for by having its path shortened. Going back to Mike and Ike, it would be as if whoever has to buck the wind would have his course shortened the exact amount necessary to compensate for the wind so that he still runs his course in the same time as his opponent, with the race always ending in a dead heat. Fitzgerald obtained the equation giving the amount of contraction necessary and, as is to be expected, it showed that the faster the ether wind, or speed of the earth through the ether, the greater was the contraction of the interferometer arm in the direction of motion. Objects moving in a direction perpendicular to the ether wind were not foreshortened, however.

The reader will immediately ask, why not just measure the lengths of the arms several times during the experiment to see if they do change? This would be impossible, since *all* objects moving with the same velocity with respect to the ether would shrink by the same fractional amount and the length of the object would always remain the same according to the measuring-tape or other length-measuring device. Nor is there any other way by which the supposed contraction can be detected.

Objections to the Fitzgerald–Lorentz contraction hypothesis were rampant, as was to be expected, not only because there was no evidence to prove that such an effect took place, but particularly because Fitzgerald could not explain why objects *would* contract due to motion through the ether. The contraction hypothesis was originally advanced only as a possible explanation for Michelson and Morley's results, *providing* such an effect existed. Then, too, the theory said that *all* materials travelling with the same velocity with respect to the ether would contract the same fractional amount. Since iron is much heavier and stronger than wood, for example, one would expect a greater contraction for wood than for iron, but this, too, went unanswered.

The contraction hypothesis was bolstered somewhat two years later in 1895, when Lorentz (hence the term Fitzgerald–Lorentz contraction) put forth his electron theory to explain the composition of matter. He postulated that matter consisted of electric charges which generated electric and magnetic fields, and that these fields resided in the ether, in keeping with the theory of the day. He reasoned that if an object were to move through the ether, it would influence the fields, due to the electric charges in the object, and cause these charges to move, thus contracting the object the same amount predicted by the Fitzgerald formula. Lorentz's theory could not be verified at that time, however, and so the Fitzgerald–Lorentz contraction could neither be proved nor disproved. We shall see in the next chapter how the theory of relativity not only verified the Fitzgerald–Lorentz contraction, but how the contraction necessarily followed as a logical consequence of the Special Theory.

The Great Dilemma

We thus see what the great dilemma was. The ether was firmly believed to exist, but all efforts to detect it not only failed but the reasons advanced for the failure were contradictory and insecure. So, did the ether exist or didn't it? If it did, why couldn't we detect it? And if it didn't exist, why didn't it?

It was at this stage of scientific frustration and confusion that the soul-satisfying answer was given, with such a simple explanation that it took a genius to see it – Albert Einstein. And with him the Theory of Relativity was born.

3 · The Special Theory of Relativity

Difference between the Special Theory and the General Theory

THE dilemma outlined in the previous chapter meant that a revolution was to take place in scientific thought if the difficulties were to be successfully surmounted. The problem was solved by what is called the theory of relativity. And it was solved in a way which not only placated all objections but which also solved other tantalizing problems not directly concerned with the problem of the ether. Nor was this all. In addition to satisfying the intellectual cravings of the scientists of the day, the theory of relativity was to make completely new and fantastic predictions which were to culminate in the dawn of the atomic age!

The theory of relativity consists of two main parts: the Special (or Restricted) Theory of relativity and the General Theory of relativity. The Special Theory was presented by Einstein in 1905 and the General Theory in 1916. In this chapter we will consider only the Special Theory, leaving the General Theory for a later chapter.

The Special Theory deals only with objects or systems which are either moving at constant velocity with respect to one another (unaccelerated systems) or which are not moving at all (with a constant velocity of zero). The General Theory treats of objects or systems which are speeding up or slowing down with respect to one another (accelerated systems). The Special Theory is really a particular case of the General Theory, since systems moving with constant velocity can be thought of as having an acceleration of zero. However, systems moving with uniform velocity were simpler to treat than those with non-uniform velocity, and so the Special Theory was arrived at first.

The Two Postulates of the Special Theory

Upon examining the large problem of the detection of the ether and the experiments which had been performed (wherein the properties of light played an important part), Einstein drew two very important conclusions. These are known as the *fundamental postulates* of the Special Theory and are the foundation which supports the rest of the theory.

The purpose in this chapter is to discuss both of these postulates in detail and then to present the results which follow if the necessary mathematics were to be carried out using these postulates as the starting-point. (The mathematical steps themselves are not included, in keeping with the purpose of this book. For the serious student, many excellent texts have been written which ably present these mathematical details, and the reader is advised to refer to them after the basic ideas and results of the theory are well in head.) The experimental proof of the various predictions of the Special Theory is convincing, indeed, and is so voluminous that the next chapter has been reserved for it.

The first postulate answered the dilemma of the ether. Stated simply, it says that *the ether cannot be detected*. But before we see why not, let us first look at some simple examples which will illustrate Einstein's reasoning in coming to this conclusion. Suppose you find yourself on a bridge over a brook which is flowing slowly by underneath. You are looking down into the water. As you gaze at your reflection in the water, it won't be long before you find it very easy to imagine that it is you and the bridge which are gliding smoothly along and that the water is perfectly calm. Of course, you don't remain in such a trance for long, because you just *know* the bridge is stationary and that it is the water which is moving.

But now we'll consider another example, where it will not be so easy to determine which of two objects is moving. Assume you are living in the future when you can hop into

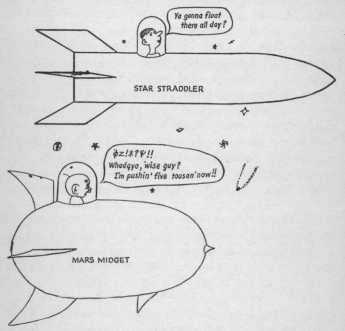

Figure 8. Which Rocket is 'Moving'?

your private rocket and take a trip out into space far beyond the earth (see *Figure 8*). You start off straight out from the earth at 5,000 miles an hour with respect to the earth, and you set your controls so you will expect to cruise through space at this speed.

After cruising until the earth is out of sight, you sight another rocket behind you. It swiftly overtakes you. You are surprised that he's going faster than you, since you had thought that yours was just about the fastest little 'hot rock' in the universe. As he passes you, you're even more surprised when he indicates that he thinks you're not even moving! But how can you prove that you are really moving? You know that he's moving at a different rate than you are because you see him approaching you. You probably would be equipped with a radar set, similar to those used by the highway police to detect violators of the speed limit, which will tell you that he's moving at 1,000 miles an hour with respect to you. But this is all you can determine.

You might think that since you left the earth, travelling at 5,000 miles an hour, and he's now passing you at 1,000 miles an hour faster, that he is then going 6,000 miles an hour with respect to the earth. But this is not necessarily true. It could also mean that you are now going 2,000 miles an hour with respect to the earth and he is going 3,000 miles an hour with respect to the earth. Or, strange as it may seem, it could even mean that he is not moving at all with respect to the earth, and that you are moving backward towards the earth at a rate of 1,000 miles an hour!

You will rapidly conclude that without some 'motionless' object to use to measure your velocity, you can never tell which one of you is moving and who is standing still, if either. You can only conclude that you are moving at 1,000 miles an hour with respect to your space friend. Nor will it ever be possible for you to develop an instrument, however complicated, which will tell you anything more but that you are moving with respect to something else. Indeed, if you are ever alone, out in space, far removed from all the stars and planets, with nothing to use as a reference point to measure your speed, you will never know whether you are moving or not!

It was this fact that Einstein recognized: *all motion is relative* (hence, the theory of *relativity*). We can never speak of absolute motion as such, but only of motion relative to something else. In general, we cannot say that an object has a velocity of such-and-such, but must say that it has a velocity of such-and-such *relative* to so-and-so. This is not done for objects on the earth because it is understood that their velocities are relative to the earth. A speed limit of fifty miles an hour, for example, is understood to mean fifty miles an hour relative to the earth. But out away from the earth a velocity by itself has no meaning.

It is easy to imagine a conversation several hundred years from now between a father and his wandering, space-lustful son. If the father tells him to keep his hot rock under 1,000 miles an hour the boy might reply in all sincerity, 'Relative to what, Daddy, the earth or the Big Dipper?' In the remainder of this book, then, we will always indicate what a velocity is relative to.

There is no heavenly body in our universe that we can use as a stationary reference point. The earth rotates on its axis; it travels in its orbit around the sun; the sun and the solar system are moving about within our galaxy, the Milky Way, which is itself rotating. And our galaxy is also moving relative to the other galaxies. The whole universe is filled with movement. And in all this seemingly haphazard turmoil, no one can say what is moving and what is stationary. We can only say that all the heavenly bodies are moving relative to one another, and no one of these is different in this respect. Nor can any one of these moving bodies be considered privileged in any way – in being the hub of the universe about which all the others revolve, for example.

But how does all this show that the ether cannot be detected? It does so as follows: A stationary ether would possess *absolute motion* because it would be the one thing in the universe which would be motionless. But we have found that we can only detect *relative motion*. Hence, we could not detect the ether. As Newton pointed out long ago, it is impossible to tell whether or not a ship is moving through the water by any experiment performed *inside the ship*. Similarly, we here on earth cannot detect the motion of the earth through an ether by any experiment performed *on the earth*.

It should be emphasized that in this first postulate Einstein does not completely reject the idea of the *existence* of the ether, but says that it can never be *detected*. Moreover, the Special Theory does not use, or need to use, the concept of the ether in any way. None of the results bears any reference to an ether. Science should not stand still, conducting a useless search for the ether. Einstein arrived at this conclusion only after many experiments had shown that the behaviour of light waves was not influenced in any way by the movement of the earth through space.

The second fundamental postulate of the Special Theory states that *the velocity of light is always constant relative to an observer*. To grasp the full meaning of this postulate, let's first investigate a situation of common experience. We will assume there's a boy throwing a baseball at 15 miles an hour. This means the ball will travel at the rate of 15 miles an hour relative to him, regardless of whether he is standing still or is on something that is moving. For example, if he is on a moving railway flat car, the ball will have a velocity of 15 miles an hour relative to him, but the speed of the ball relative to the ground will be more or less, depending on the car's velocity and direction of motion. In particular, if the car is moving towards a bridge at 5 miles an hour, as in *Figure 9(a)*, and the boy throws the ball towards the bridge, the speed of the car is added to that of the ball and the ball travels over the ground at 20 miles an hour, hitting the bridge with a 20-mile-an-hour impact. Conversely, when the boy is travelling away from the bridge, as in (*b*), and throws the ball at it, the speed of the car subtracts from that of the ball, which hits the bridge with only a 10-mile-an-hour impact.

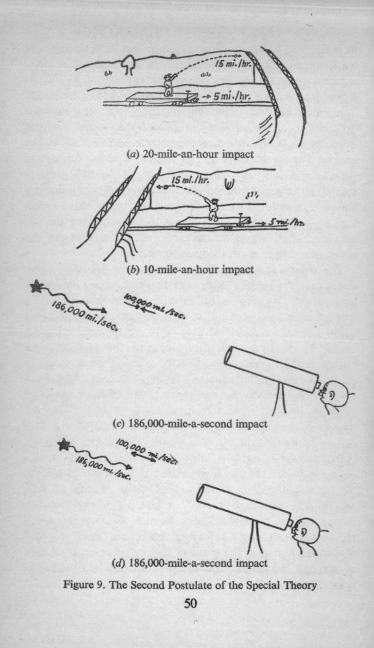

(a) 20-mile-an-hour impact

(b) 10-mile-an-hour impact

(c) 186,000-mile-a-second impact

(d) 186,000-mile-a-second impact

Figure 9. The Second Postulate of the Special Theory

50

Now replace the boy by a distant star, the bridge by a telescope on the earth, and the thrown ball by a light wave travelling through space from the star to the telescope. The light wave, then, will be 'thrown' from the star with the speed of light, i.e., 186,000 miles a second relative to the star. But here the similarity between the two situations ends. If the star and earth are approaching each other with a relative velocity of 100,000 miles a second, as in (c), we would expect this velocity to be added to that of the light wave in the same way that the velocity of the moving flat car was added to that of the thrown ball. This means that the light wave should produce a 286,000-mile-a-second 'impact' on the observer's eye. Conversely, if the star and earth were separating from each other with a relative velocity of 100,000 miles a second, as in (d), we would expect the velocities to subtract and produce an 86,000-mile-a-second impact on the observer's eye.

Comparing both of these results, we see that the velocity of light is expected to be *different* in each case. But this would contradict the second postulate, which states that the velocity of light is always the *same* relative to an observer. Hence, the above reasoning cannot be true, i.e., light cannot have different velocities relative to an observer. The only conclusion, then, is that the light waves will produce a 186,000-mile-a-second impact on the observer's eye in both cases. And, according to the second postulate, it makes no difference how fast or slow an observer and a light source are approaching or separating; the velocity of light will *always* be constant at 186,000 miles a second relative to the observer.

Just think of it. This means that a light wave leaving a star will have a velocity of 186,000 miles a second relative to an observer, regardless of whether he and the star are approaching or separating at a relative velocity of 185,999 miles a second or 1 mile a second! If this same postulate governing light waves were to apply to the boy on the flat car, it would mean that (still assuming the boy throws the ball at 15 miles an hour) the ball would always hit the bridge at the same speed (15 miles an hour), regardless of how fast or slow the flat car was moving towards or away from the bridge!

This postulate was a revolutionary statement. Yet, Einstein made this one of his basic postulates – even though it seemed to defy common sense – since all the experiments pointed to this conclusion. He believed it to be one of the basic laws of the universe.

Deductions from the Postulates

Since the two postulates were so contrary to common thought at the time, they needed much more than mere public presentation. For, without further support, they would have been regarded only as interesting and would not have proved anything. So, using these postulates as a starting-point, a number of equations were derived which not only explained particular phenomena but also made certain predictions which were later verified by experiment. This is the really stringent test of any theory – not only satisfactorily to explain all the puzzles of a problem, but to make entirely new and different predictions which experiment can later verify.

In order to bridge the gap between the postulates, seemingly abstract in themselves, and the equations leading to the verification and practical applications of the theory, the postulates have to be incorporated into a physical situation susceptible to experiment. Since the postulates deal with an object moving at a constant velocity relative to an observer and the behaviour of light waves, this can best be done by having an observer 'describe' the object while it is moving at a constant velocity relative to him. The peculiar behaviour of light waves will markedly influence the description, since it is the reflection of the light waves from the object to the observer which enables him to see and to describe the object. The observer's 'description' of the object will consist of its physical characteristics which are actually measured by the observer's instruments, such as its length, mass, etc. The prediction of the numerical values of these quantities according to the Special Theory are put in mathematical form to enable them to be compared with the actual measurements. The method for deducing these mathematical predictions of the Special Theory will now be indicated briefly.

Assume we have two identical rockets, A and B, which are travelling with a finite velocity relative to each other out in space (*Figure 10*), A and B are each equipped with at least the most elementary scientific instruments, particularly measuring-sticks and clocks, and it is especially important that these be compared beforehand so that A's be identical with B's. When the analysis begins, B is passing A, both their clocks read the same time, and a nearby supernova explodes at the same instant. Neither A nor B is aware that the star has exploded, since the light waves of the explosion haven't reached them yet.

A short time later, the light waves from the explosion reach A and B, but when they do, A and B will be separated by the distance x. From the second postulate, A and B see the light waves coming in with the same velocity relative to each and, letting c represent the velocity of light waves for A and c' that for B, we can say that $c = c'$. The distance of the explosion from each, d and d', and the times given by each of their clocks, t and t', are then incorporated, and the analysis proceeds to interrelate their distance from each other, their relative velocity, their respective times, the velocity of light, etc.

The resulting equations are called the *Lorentz transformation equations* because Lorentz had previously arrived at the same equations on the basis of his theory. However, his theory was artificial, being based on the necessary existence of the ether, and was not logically consistent. (Review, for example, the reasoning which led to Lorentz's contraction hypothesis.) Then, too, some of his results applied *only* to electric and magnetic fields. The Special Theory of relativity, on the other hand, rests solidly on the two fundamental

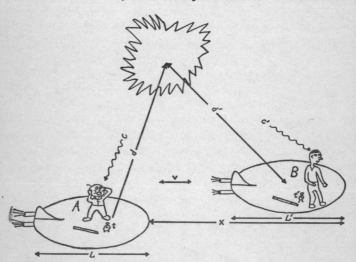

Figure 10. The Basis for the Lorentz Transformation Equations

postulates, and their results apply to *all* matter without exception.

Using the Lorentz transformation equations, we can now predict the result each obtains if he scrutinizes the other closely as to his length, mass, etc. We now proceed to discuss each of these in detail. And since the postulates involved conclusions contrary to everyday experience and are the basis for this analysis, the reader should not be surprised to find that the results will also be unexpected and seemingly queer. The reason the theory of relativity is in general looked upon as being incomprehensible is not because the results are difficult to understand but that they are difficult to believe.

Contraction of Length

If A is able to measure B's length when they are moving with velocity v relative to each other, the mathematical results predict that B appears to have shrunk, his length being given by *Equation 1*:

$$L' = L\sqrt{1 - \frac{v^2}{c^2}}$$

where L' is the length A obtains for B, L is B's original length, v their relative velocity, and c the velocity of light. As an illustration, if A and B were each of length 20 feet when at rest with respect to each other but are now separating at a relative velocity of 93,000 miles a second (half the velocity of light), then B's apparent length as measured by A can be determined by substituting these values in the equation, with the result that A would measure B to be only 17 feet long. Or, if they were separating at 161,000 miles a second (about nine-tenths the velocity of light), then, to A, B's length would appear to be only 10 feet. Also, since we said that the rockets were identical, their length should be equal if they are not moving with respect to each other at all (relative velocity of zero), and the equation should verify this. It is seen that it does, since when v equals zero, the value of the radical is one and $L' = L$. Hence, with B at rest with respect to A, A would find B to be 20 feet long.

Now the reader asks, what value does B get if he measures A's length in passing? In this case the formula still applies, but now L' and L should be interchanged, since L' in reality is the length seen by the observer doing the measuring. Here, the results are the same, i.e., at a separation velocity of 93,000 miles a second A's rocket will appear to be 17 feet long to B, and at 161,000 miles a second it will appear to be only 10 feet long. And with A and B at rest with respect to each other, B will find A to be 20 feet long. Nor does it matter whether they are separating or approaching – the result is still the same, depending only on their *relative* velocity.

Now, what if A measures his own length while B is passing? If he does, he would find it to be 20 feet long, since he is not moving with respect to himself. Of course, it is immaterial whether B is passing at the same time or whether A is moving relative to *any* other system. A *always* obtains 20 feet for his own length. Similarly, if B measures his own length while moving or not with respect to A or any other system, he also *always* obtains 20 feet.

This effect of length contraction can be stated simply: whenever one observer is moving with respect to another, whether approaching or separating, it appears to both observers that everything about the other has shrunk in the direction of motion. Neither observer notices any effect in his own system, however.

It is seen that the contraction effect is appreciable only if the relative velocity is comparable to the velocity of light. Since the velocities with which we are familiar on earth are considerably lower than the velocity of light, we do not ordinarily notice the contraction effect. For example, *Equation 1* shows that an aeroplane moving at a rate of 750 miles an hour relative to an observer would shrink by about a millionth of a millionth of an inch, or about the diameter of the nucleus. Such small amounts are not even detectable with our most precise instruments – let alone being noticed by the human eye.

It may seem to some readers that the preceding discussion is very artificial and hence, invalid, inasmuch as it would be well nigh impossible actually to measure the length of a rocket with a ruler as it passes at a velocity of 93,000 miles a second. Then are the conclusions predicted by *Equation 1* without meaning? The answer is that the conclusions *are* valid. Although a ruler was used for illustration because it is the simplest device for measuring length, the results would apply *regardless* of how the lengths are measured. In an actual experiment the measuring equipment would be quite complicated, involving electronic circuits, light beams, etc.

For historical reasons, the contraction effect is still referred to as the Fitzgerald–Lorentz contraction and is adequately expressed in the now-famous limerick:

> *There was a young fellow named Fisk*
> *Whose fencing was exceedingly brisk;*
> *So fast was his action,*
> *The Fitzgerald contraction,*
> *Reduced his rapier to a disk.*

Mass Increase with Velocity

The next important result we consider is the increase in mass with velocity. Suppose that A and B each have a mass on earth of 1,000 lb when at rest with respect to each other. Now, if A measures B's mass while they are moving relative to each other, he will find that B's mass appears to have increased, its numerical value being given by *Equation 2*:

$$m' = \frac{m}{\sqrt{1 - \frac{v^2}{c^2}}}$$

where m' is the value A obtains for B's mass, m is B's original, or rest mass, as it is called, v their relative velocity, and c the velocity of light. This means that if A and B both have a rest mass of 1,000 lb each when at rest with respect to each other on earth, then when they are approaching or separating at a relative velocity of 93,000 miles a second, *Equation 2* shows that B would appear to have a mass of about 1,200 lb if A measures B's mass by attempting to stop him or by some other similar method. At 161,000 miles a second, B's mass would be 2,000 lb, or twice as great! And at higher velocities, B's mass would be even greater, the exact amount being given by the equation.

If B measures A's mass, he, too, finds that A's mass will appear to have increased to the amount given by *Equation 2*. Now, however, m' will be the mass B obtains for A, and m is A's rest mass (which, of course, is still the same as B's rest mass).

If A and B measure each other's mass when they are at rest with respect to each other, then v is zero in *Equation 2*, the value of the radical is one, and m' equals m, i.e., both their masses are equal and still 1,000 lb each, just as we expect. Further, each will *always* determine his own mass to be 1,000 lb, regardless of how he is moving relative to anyone else, since he does not move relative to himself.

The mass-increase equation, then, states that when an object is moving with respect to an observer, the mass of the object becomes greater, the amount of increase depending on the relative velocity of object and observer.

It is ironic that some corpulent people attempt to decrease their mass with vigorous exercise, often by running. But the Special Theory of relativity says that their mass will *increase* – the faster they run, the greater their mass becomes! To cite a numerical example, if a 300-lb man runs at 15 miles an hour (entirely unlikely), his mass will increase by about a millionth of a millionth of an ounce, or 0·000000000001 ounce! (Of course, the effect would be greater if he could run faster.)

An analogy illustrating the mass increase with velocity occurs with a ship moving through the ocean. Some water is always dragged along behind the ship. And the faster the ship goes, the greater is the amount of water dragged along. The ship thus effectively acts as if its mass increases as its velocity increases, since the water being dragged along moves with the ship and becomes part of the ship's bulk.

The reader is cautioned against believing that the mass-increase effect means an object becomes bigger in the sense that its physical dimensions (length, width, height) increase. This is not true. You can envision an object becoming heavier without it becoming larger. Remember, the contraction effect predicts that it actually becomes smaller in size while its mass increases, when moving relative to the observer.

Addition of Velocities

To illustrate what the Special Theory predicts for the addition of velocities, consider *Figure 11(a)*, which shows two automobiles, A and B, approaching a pedestrian, each with a velocity of 100 miles an hour relative to the pedestrian. This means that if the pedestrian measures the velocity of each car, he will find it to be 100 miles an hour. Or, conversely, if each driver measures his velocity relative to the pedestrian, each one obtains 100 miles an hour. If A measures his velocity relative to B, he will find it to be 200 miles an hour, since they are each doing 100 miles an hour relative to the pedestrian. Similarly, B finds that his velocity relative to A is 200 miles an hour. For this general situation we customarily use *Equation 3*:

$$V_{AB} = V_A + V_B$$

where V_{AB} is the velocity of A or B relative to each other, V_A is A's velocity relative to the pedestrian and similarly for V_B. But now suppose we consider a situation where the velocities are considerably higher, as in (b). Assume that A and B are in rockets out in space, each travelling towards a 'spacetrian' with velocities relative to him of 100,000 miles a second each. If the spacetrian measured A and B's velocities relative to him, he would determine them each to be 100,000 miles a second. And similarly, A and B would each determine his own velocity to be 100,000 miles a second relative to the spacetrian.

But now if A and B each measures his own velocity relative to the other, the Special Theory says that they will not each obtain 200,000 miles a second, as would be given by

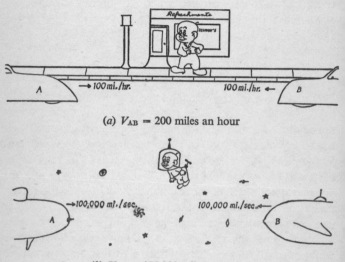

(a) $V_{AB} = 200$ miles an hour

(b) $V_{AB} = 155,000$ miles a second

Figure 11. The Addition of Velocities

Equation 3. Rather, the theory says that their relative velocity is given by *Equation 4*:

$$V_{AB} = \frac{V_A + V_B}{1 + \frac{V_A V_B}{c^2}}$$

where V_A and V_B are the relative velocities of A and B with respect to the spacetrian, and c the velocity of light. If we insert the above values for V_A and V_B and 186,000 miles a second for c, we see that the relative velocity of A and B is only 155,000 miles a second!

If *Equation 4* is correct, then *Equation 3* is not. For all practical purposes, however, *Equation 3* can be considered correct for velocities considerably lower than the velocity of light. For example, in the case of the two automobiles approaching each other with a velocity of 100 miles an hour each relative to the pedestrian, *Equation 4* predicts that their *exact* relative velocity is about a millionth of an inch *less* than 200 miles an hour.

The Maximum Possible Velocity

Perhaps the most astonishing prediction to come out of the Special Theory is that there is a certain velocity beyond which nothing can go. To see what this is, we refer to *Equation 1*, which governs the contraction of an object with velocity. We found that the faster an object travelled relative to an observer, the shorter it became. We now ask, what happens if we increase the velocity more and more? Does the object disappear? This is precisely what the formula says *does* happen, for it is easily seen that as v approaches the velocity of light, c, the length of the object approaches zero. And at v equal to c, the length *is* zero, which means that the object *has* disappeared!

But now suppose we make v *greater* than c in the equation, say equal to twice the velocity of light, $2c$. Then we will get the negative number 3 under the radical. This tells us that the length of the object is now its original length times the square root of minus 3. But mathematicians tell us we cannot take the square root of a negative number – such a number is purely imaginary. Hence, in this case the length of the object will be imaginary and the object itself will not exist.

Let's see what *Equation 2* predicts for the mass of an ob-

RELATIVITY FOR THE LAYMAN

ject as its velocity approaches that of light. As *v* increases, the radical in the denominator decreases and, since the value of a fraction increases as the denominator decreases, the mass of the object will increase, as we found previously. Furthermore, if we let *v* increase to the point where it is equal to the velocity of light, *c*, the denominator becomes zero, which means that the mass becomes *infinite*.

The only conclusion that can be drawn from all this is that *the velocity of light is the maximum possible velocity*. Nothing can travel faster than light because, as we have seen, not only does its length shrink to nothing but its mass becomes infinite. And, as a matter of fact, it is more correct to say that the material objects with which we are familiar can never even travel as fast as light, because their mass would become infinite, which means that an *infinite* amount of energy would be required to get them up there. An infinite amount of energy means *all* the energy in the universe *plus a great deal more*.

It can now be seen why *Equation 4* does not give a value of 200,000 miles a second for the relative velocity of A and B in *Figure 11(b)*, as we first expected from *Equation 3*. Two hundred thousand miles a second would be *faster* than the speed of light, and this would be impossible. Regardless of how fast two objects move with respect to an observer, they will find that their velocity relative to each other is *always less* than the speed of light. For example, suppose that both A and B in *Figure 11(b)* had relative velocities of $0.9c$ with respect to the spacetrian. Substituting these values in *Equation 4*, it is seen that, if they measure their velocity relative to each other, they will each get $0.99c$, which is less than the velocity of light.

From a philosophical point of view the following argument could be raised. Suppose a large rocket ship is fired straight out from the earth at a velocity of $0.9c$ relative to the earth. When the rocket is well on its way, it in turn fires a small projectile straight out from its nose at a velocity of 0·9c relative to the rocket. Surely, the philosopher argues, the projectile must actually be moving at a velocity of 1·8c relative to the earth. You, he continues, are saying that the velocity only *appears* to be 0·99c relative to the earth. The scientist's answer is that he can believe only what his instruments tell him. Here, his instruments and observations would tell him that the projectile moves at a velocity of 0·99c relative to the observer on the earth, since they are governed by the physical laws of the behaviour of light waves. This would constitute *proof* for the scientist, and no one could furnish contradictory proof.

The surprising results set forth in this section are emphasized in the famous limerick:

> *There once was a lady called Bright,*
> *Who could travel faster than light;*
> *She went out one day,*
> *In a relative way,*
> *And came back the previous night.*

Equivalence of Mass and Energy

The one particular result of the Special Theory of relativity which has had a most far-reaching effect on our age is the prediction that a comparatively small amount of matter is equivalent to an enormous amount of energy. As everyone now knows, the first convincing example of this occurred when the first atomic bomb was exploded at Alamogordo, New Mexico, on 16 July 1945. (The theory of the atomic and hydrogen bombs will be discussed in the next chapter.)

The formula relating mass to energy is arrived at as follows: We have found that the mass of an object increases with its velocity. Now, it follows that its energy must also increase, since a heavier object has more energy than a lighter object, providing they both have the same velocity.

It can be shown that the additional energy associated with the additional mass is equal to the increase in mass times the velocity of light squared. And since it is found that the *additional* mass has an associated energy, why can we not assume, then, that *all* of *any* mass has an associated, or equivalent, energy where this energy is given by the product of the mass and the square of the velocity of light? This is the conclusion which Einstein drew. Stated mathematically (*Equation 5*):

$$E = mc^2$$

where E is the equivalent energy, m the mass of the object, and c the velocity of light.

The interpretation of *Equation 5* is that if a mass of *any* substance is converted to energy, without any mass being left over, the amount of energy obtained is given by this equation. As an example, if the mass of a pound of coal is inserted in the equation and the appropriate units used, the reader easily can verify for himself that the equivalent energy is about 30,000,000,000,000,000 foot pounds! This is about the same as the total amount of energy currently generated by all the power-stations in the United States in one month! Even a teaspoon of coal dust would supply enough energy in this way to drive our largest ocean liner from New York City to Europe and back many times.

The reader is perhaps wondering what happens when we burn a pound of coal in the ordinary way. Isn't energy released then? Of course it is, but here the process is a purely chemical one – the molecules are rearranged in combining with the oxygen molecules in the air, and in doing so heat energy is released. But no *measurable* conversion of mass to energy takes place, because the coal changes into soot, ashes, gases, etc. If these end products are weighed, their combined weights still will add up to the original 1 lb approximately. If you compare the amount of energy released

64

by a pound of coal when burned with that released by completely converting its mass to energy, you will find that the latter will produce three billion times more energy. Of course, the process whereby an appreciable amount of mass is converted to energy is completely different from ordinary burning. (These so-called *nuclear processes* will be discussed in the next chapter.)

Time in the Special Theory

So far we have not said anything about how A and B's clocks compare in *Figure 10*. We assumed their clocks were identical and that they both read the same time at the instant A and B were alongside each other. We will let the time be 12 o'clock. For mathematical reasons we will call this time the zero hour.

A short time later, A and B are separated by a distance x. If A then looks at his clock and compares its reading with B's, he will be surprised that B's clock appears to be running slower. This is precisely what the Special Theory predicts, since mathematical results show that the times given by the clocks are related in this particular case by *Equation 6*:

$$t' = t\sqrt{1 - \frac{v^2}{c^2}}$$

where t' is the time A reads for B's clock, and t the time A reads on his own clock. As an example, if we assume that A and B's relative velocity is 93,000 miles a second, then B's clock will appear to A to be running about nine-tenths as fast as A's. If A finds that his own clock reads 1 hour, i.e., 1 o'clock, at a particular instant, *Equation 6* shows that t', B's time as read by A, is about 54 minutes, i.e., only 6 minutes to one. And regardless of what time A looks at his clock, he will always find that B's clock reads only nine-tenths as much.

If their relative velocity were 161,000 miles a second, the equation would show that B's clock would appear to A to be going only half as fast. Now, when A finds it is 1 o'clock on his clock, he finds that B's clock is lacking half an hour of 1 o'clock. And the higher their relative velocity, the slower B's clock appears to go. Nor does it make any difference whether A and B are approaching or separating – B's clock will always appear slow to A.

If B reads his own clock and compares it with A's, it will appear to him that A's clock is slow, for now t in the equation refers to B's reading of his own clock and the t' to B's readings of A's clock. At the relative velocity of 93,000 miles a second, A's clock will appear to B to be only about nine-tenths as fast. Similarly, at 161,000 miles a second, A's clock will appear to B to be only half as fast.

These results are expected, since we have found that when two observers are moving relative to each other, any effects that occur are the same for each. Of course, if two observers are not moving relative to each other, i.e., if their relative velocities are zero, then t' equals t, and both clocks would read the same time, as we would expect.

This *time-dilation effect*, then, states that if two observers are moving at a constant velocity relative to each other, it appears to each that the other's time processes are slowed down.

You can rightly conclude from the foregoing discussion that the reason A and B's clocks appear slow to each other is not only because of the peculiar behaviour of light waves, as set forth in the postulates, but because it also takes a certain interval of time for the light waves to travel from one to the other. This time-dilatation effect was responsible for an entirely different outlook regarding time from what had been in vogue previously. Time had always been considered to be the same for everyone; that is to say, time passed at the same rate for every person or object in the universe. Like a large, slow-flowing river whose current is the same for all points along its banks, time was considered as a uniformly flowing thing which passed at the same rate for everyone. The Special Theory showed that this was not true. It showed that time flowed at *different* rates for two observers moving relative to each other.

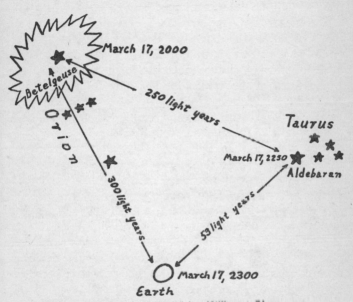

Figure 12. Time is Different for Different Observers

We now examine a simple case which shows that time is different for different observers at different positions *not* moving relative to each other. To be strictly correct, we should say that in this case the dates (meaning fixed points in time) are different, since the rate of time processes are the same for each, as *Equation 6* verifies, since the relative velocity, v, is zero here. To illustrate, consider *Figure 12*, which shows the earth, the star Betelgeuse in the constellation

Orion, the Hunter, and Aldebaran in Taurus, the Bull. Betelgeuse and Aldebaran are 300 and 53 light years, respectively, from the earth. Also, Aldebaran is about 250 light years from Betelgeuse.

Now suppose there is a 'blow-out' in Orion on the night of March 17, 2000, caused by Betelgeuse exploding. This date, and all the dates mentioned here, refer to our method of keeping track of time on earth. We on earth would not see the blow-out on that date, since Betelgeuse is 300 light years away, which means it would take 300 years for the light waves of the explosion to reach us. This is the only way we would learn of the explosion. The date for us would be 17 March 2300. Somebody on Aldebaran on the other hand, would see the explosion on 17 March 2250, since Aldebaran is 250 light years from Betelgeuse.

It can be seen, then, that the *single event* of the explosion is not *simultaneous* to the three different places, since each observes the event at a *different* time. Of course, we on the earth could, theoretically, compare our observation of the explosion with that of observers on Aldebaran and agree on just when, in reference to our individual time-recording methods, the explosion did occur on Betelgeuse assuming all observers agreed on their distances and relative bearings from each other. But before the theory of relativity was introduced, the distance between two different positions was determined merely by laying off the distance with a measuring-tape or other suitable device. Time never entered into the measurement, because it was considered to be the *same* at the two different positions. However, we have just shown that this is not true; time is *different* at two different positions. So, strictly speaking, recognition must be taken of this fact by including time in the measurements.

Figure 13 illustrates the equations for determining the distance between two points for an increasing number of dimensions. In one dimension the length *OA* is merely the

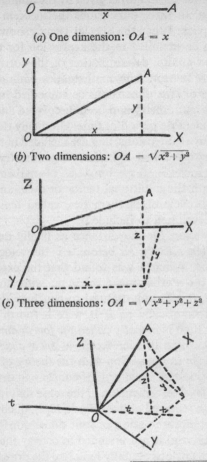

(a) One dimension: $OA = x$

(b) Two dimensions: $OA = \sqrt{x^2 + y^2}$

(c) Three dimensions: $OA = \sqrt{x^2 + y^2 + z^2}$

(d) Four dimensions: $OA = \sqrt{x^2 + y^2 + z^2 - (ct)^2}$

Figure 13. Determining Distance Between Two Points for
One, Two, Three, and Four Dimensions

69

distance along the x axis, and the measurement is trivial. In two dimensions the length is given by the familiar Pythagorean theorem. In three dimensions the theorem is extended and still applied. When the Special Theory showed that time also had to be included in the expression for the distance between two points, determination of the correct equation was no easy matter. The mathematics embracing all the known laws of two dimensions as embodied in plane geometry and trigonometry had been discovered and developed over a considerable period of time. Gradually these were extended to three dimensions, and are included in the branches of mathematics called spherical trigonometry and solid geometry. However, these branches of mathematics could not cope with the additional factor of time, and so an entirely new branch of mathematics, called *tensor calculus*, had to be developed to include it.

The expression for the distance as finally determined is shown in *Figure 13(d)*. As before, c is the velocity of light, and t is time. When it was found that the expression was similar to the Pythagorean theorem, with an additional factor, $(ct)^2$, it was only natural to conclude that time appeared mathematically *as if* it were a fourth dimension. This is why time is *actually called the fourth dimension*, and explains the origin of such words as *space-time* and *space-time continuum* in connexion with the theory of relativity.

The reader is cautioned against concluding that time is an additional physical dimension in the sense that it can be seen and felt like a material object. *Figure 13(d)*, of course, is only the author's representation of four dimensions for illustrative purposes, and is not intended to convey the impression that four dimensions actually look like the drawing. No one in our universe can see in four dimensions or more because of the way our universe is constructed. Some people insist they can 'think in five (or more) dimensions' – whatever this means.

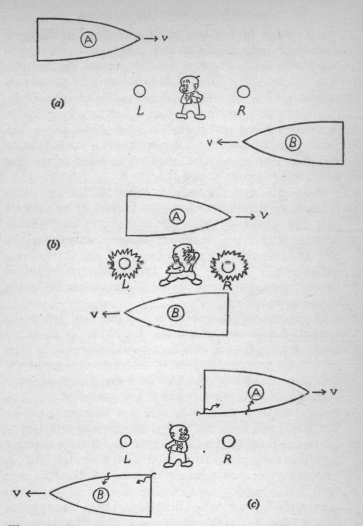

Figure 14. Two Events which are Simultaneous for an Observer 'at rest' are not necessarily Simultaneous for Two Other Observers not 'at rest'

The change in the *rate* of flow of time for different observers moving relative to each other also affects the so-called *simultaneity* of events. Einstein showed, on the basis of the Special Theory, that while two events might be simultaneous for one observer, they are not necessarily simultaneous for all observers. In fact, it can very easily occur that two 'happenings' could be simultaneous for a first observer, but for a second observer one event would precede the other. And for a third observer the order of the two events could even be reversed!

This possibility is illustrated in *Figure 14*. Two observers, *A* and *B*, are in identical rockets and are travelling towards each other. They each have a velocity *v*, slightly less than that of light, relative to a 'stationary' observer, who is situated halfway between two explosive charges, *L* and *R*, as shown in *Figure 14(a)*. The charges are at rest relative to the stationary observer and separated by a distance equal to the length of one of the rockets.

The situation of the rockets is such that they pass at the charges with the nose of *A* lining up with charge *R* and the tail of *B*, while the tail of *A* lines up with charge *L* and the nose of *B*'s rocket. The charges are set to explode at the instant the rockets do pass each other (see *Figure 14(b)*).

Assuming that both *A* and *B* are located at the geometrical centre of their rockets, we now examine the order in which *A* and *B* see the explosions. Observer *A* moves to the right a short distance during the time it takes the light rays from the explosions to travel to *A*, which tell him the 'happening' of the explosions has occurred. The light rays from *R* thus have a slightly shorter distance to travel and will reach *A* first, i.e. before the light rays from the explosion at *L*. *A* thus sees the light rays from *R* first, and he therefore concludes that *R* exploded before *L* (see *Figure 14(c)*).

With *B* the situation is reversed. He is moving to the left closer to *L*, while the light rays from both explosions are moving towards him and the light rays from L reach him before those from R. B therefore concludes that charge L exploded first.

Here we have a situation where two separate events which are simultaneous for the stationary observer are not simultaneous for the other two observers. In fact, from A's point of view R exploded before L, but from B's point of view L exploded before R. No one can say which 'really' exploded first, or if they were simultaneous, because all three observers are equally correct, and no one of the three's views is to be preferred over the others. The age-old idea of the simultaneity of events – i.e. that if two events were simultaneous for one observer, they were simultaneous for all observers – was shown to be invalid by the Special Theory of relativity. The order of events is a function of the observer's position and velocity relative to all other observers. Simultaneity is a relative matter.

The two events of the two explosions described above were separated in space by a distance equal to the length of one of the rockets. As the distance between two events becomes greater, the greater is the possible difference in time between the events as seen by different observers under different conditions. Conversely, when the distance between two 'simultaneous' events decreases to the vanishing point, i.e. when two 'simultaneous' events occur at the same place, all observers, regardless of the circumstances of their positions and relative velocities, agree on the simultaneity of the two events.

For example, if two rockets collided in space, all observers would see the event of both rockets being crushed as a single event. It would not only be ludicrous but also a gross violation of the basic laws of physics if any observer would observe that one rocket was crushed, without a physical cause, before the other.

The fact that time appears slower on the other's system to

two observers travelling with a constant velocity relative to each has some interesting applications to space travel. To illustrate, we must first make clear that when time appears slower on a moving system, not only do the clocks on that system appear slower but *all* time processes are slowed down. This means the digestive processes, biological processes, atomic activity – all are slowed down.

With this in mind, then, we assume that a rocket ship, manned by several men, makes a space trip to Arcturus in the constellation Bootes, the Herdsman, which is thirty-three light years from the earth. If the rocket travels at a velocity close to the velocity of light, it will arrive on Arcturus a little over thirty-three years later, earth time. And if it returns immediately, it will arrive back on earth about sixty-six years after leaving the earth.

Since the rocket has been moving with a high velocity relative to the earth, *all* time processes in the rocket have been slowed down considerably. It will not seem to the men on the rocket that it took thirty-three years to make the one-way trip. In all probability they will be pulling into the region of Arcturus just about the time they get gurgles in their stomachs to tell them they're hungry. And when they arrive back on earth it will seem to them that only a day has elapsed. But to the people on earth it will have been sixty-six years. When the men alight from the rocket they will find that their wives, who were young when they left them, are now too old and feeble to come and meet them or, still worse, that they have long since died of old age! And some of the men might even be faced with the startling prospect of having to greet a hitherto unknown son or daughter who is sixty-six years older than they! It might seem enticing to keep young by travelling about in space, but it does have its complications!

One of the results of the time prediction which has been a source of great puzzlement and some disagreement since its introduction is the *clock paradox*, also known as the *twin*, or *time paradox*. It can be explained best by an example.

We assume we have twins, one of whom makes a trip to a distant star and back while the other remains on the earth. We further assume the star to be at a distance of four light years from the earth and that the rocket twin travels at an average velocity of four-fifths that of light. The total time for his trip, then, will be about ten years, earth time.

The question of interest is how fast time flows for the rocket twin as compared to the earth twin for the trip as a whole. Inserting for *v* the value four-fifths that of light in *Equation 6*, we find that

$$t' = \frac{3t}{5}$$

This says that, on the average, time for the rocket twin moves only three-fifths as fast as it does for the earth twin. This means that while the total trip took ten years according to the earth twin's clock, it took only six years according to the rocket twin's clock. Thus, when the twins are reunited again on the earth the rocket twin will find that he has not aged as much as his stay-at-home brother. In fact, the rocket twin will find that he is actually four years younger, biologically.

Now for the paradox. Since all motion is relative, why can't we assume that it is the earth which moves off into space in a direction opposite to that of the rocket originally and returns while the rocket ship remains at 'rest'? The earth would travel a total distance of eight light years at an average velocity of about four-fifths that of light just as the rocket did. This appears to be justifiable mathematically, because in *Equation 6* the change would be to replace *v* by -*v*. The result would not be changed, however, because *v* is squared in the equation.

If we accept this opposite view of the entire situation, then the mathematical prediction from the time dilation formula is reversed. Now the earth twin will be four years younger than the rocket twin when they get together again.

So the paradox is that when the rocket twin takes this particular trip he returns four years younger than his earth twin, but if we assume it is the earth which moved off into space and returned, it is the earth twin who returns physically younger than the rocket twin. Obviously, each cannot be younger than the other simultaneously, so we have a contradiction. This constitutes the paradox.

The solution to the paradox is that both views are really not interchangeable, as we have assumed, so there is no paradox at all. In other words, the two situations are not symmetrical, and therefore not reversible mathematically. The reason is that the rocket undergoes accelerations and decelerations during the trip, whereas the earth does not. The assumption that the earth goes out into space and returns, rather than the rocket, is not valid, because the earth would have to undergo the necessary accelerations and decelerations instead of the rocket, which we know it does not do.

Of course, if the earth were to actually make the trip by undergoing the necessary accelerations and decelerations while the rocket remained at 'rest', then, in this most improbable case, it would be the earth twin who would be the younger, physically, when they were reunited. It is also possible, too, to envision the perfectly symmetrical case wherein the time dilation effect would be the same for both twins so that they would each age at the same rate. This would occur if the twins had identical rockets initially at 'rest' at the same point on the earth and then travelled off into space in opposite directions. We assume that their accelerations, decelerations, average velocity and distance travelled are the same for each. Now, when they returned to earth they would each be the same age physically, i.e. they will have both aged the same amount. For the particular figures cited above the total travelling time for each twin, according to his own clock, would be six years. An observer on the earth, however, would record the total time to be ten years.

Returning to our original case now, the rocket could not avoid the deceleration and acceleration necessary at the distant star to reverse its velocity by turning in a large semi-circle. In order to turn in a circular arc a sideways force must be imparted which results in an acceleration at right angles to the direction of motion. Thus, the use of some type of acceleration to reverse direction and velocity is always necessary.

A rigorous mathematical treatment of the situation using *Equation 6* plus the other formulae which apply under the conditions of the Special Theory leads to the inescapable conclusion that less total time for the round trip will elapse for the space traveller, regardless of how he measures time, than for 'Earthians'. Any future 'spacetrian' will always return to earth at the end of his trip not having aged as much as Earthians who stayed behind. The total amount of the time retardation predicted by the Special Theory for the spacetrian will depend, of course, on the velocity of the rocket relative to the earth, assumed constant, and the total distance of the trip.

The physical basis for the foregoing conclusion is seen by comparing what each twin sees when each observes the light waves received from events occurring in the other's system. Consider first the observations the rocket twin makes of events on the earth. During the outgoing half of the trip, because of its velocity of recession away from the earth, the light waves of the events on earth will overtake the rocket at a slower rate than if the rocket were at rest on the earth. For the rocket's velocity of four-fifths that of light this rate is given by the so-called *relativistic Doppler shift* formula as one-third the normal, or 'rest', rate. Similarly, during the return trip the rocket twin observes events on the earth as occurring at three times the normal rate. During the entire trip, then, the rocket twin records earth events as occurring at an average rate of five-thirds (the average of one-third and three) the normal rate for him. The result is that the rocket twin notes time on the earth to be moving faster, on the average, than on the rocket with the precise ratio being five-thirds. Thus, ten years on the earth would correspond to only three-fifths of this, or six years, on the rocket.

From the point of view of the earth twin the situation is reversed. Under the same physical conditions being considered here the earth twin will receive the light waves from events on the outgoing rocket for a total of nine years. This is because it will take the rocket five years, earth time, to reach the distant star plus an additional four years for the light waves of this event to come back to the earth, since the star is four light years from the earth. And during these nine years the earth twin observes the events on the rocket to be occurring at one-third the normal rate for him on the earth, according to the relativistic Doppler shift formula. Thus, only three years of events on the rocket are observed by the earth twin during these nine years.

Events on the rocket during the return trip are seen by the earth twin to occur only during the tenth year of the trip, since the total trip is ten years, and only events of the outgoing trip have been observed during the first nine years. During this last year events on the rocket are observed by the earth twin at three times the normal rate for him, again according to the relativistic Doppler shift formula. The earth twin thus records three more years of rocket events during the final year of earth time. The net result is that six years of events on the rocket correspond to ten years of events on the earth. This agrees with the previous conclusion. In other words, time does move slower, on the average, on the rocket than on the earth.

The above discussion also indicates why the physical situation is not symmetrical for both twins and why the total time for the trip is different for each. The rocket twin reverses his velocity at the half-time of his trip and begins to perceive events happening on the earth at the increased rate immediately thereafter, whereas the earth twin must wait an additional length of time (four years in this case) for the light waves of the 'turning-around' event to reach him before he receives the light waves from the returning rocket at the increased rate. More simply, the earth twin receives the light waves of events from the rocket at a slower rate for a longer time (four years longer) than the rocket twin does those from the earth. The net effect of this asymmetry is that the earth twin observes fewer events occurring on the rocket than the rocket twin observes of events occurring on the earth during the total round trip time for each.

Now, the 'events' which each observes could very well be the number of ticks of identical synchronized clocks they were each equipped with before the start of the trip. The rocket twin would record fewer ticks on his clock than the earth twin would on his, thus indicating that less time elapsed on the rocket than on the earth during the trip as a whole. And since the ageing process is directly related to time, the rocket twin will have aged less, i.e. be biologically younger, than his stay-at-home brother by the end of the trip.

It might seem as though these conclusions contradict the prediction from the Special Theory that the velocity of light is the limiting velocity for all material objects. If we take the round-trip distance to the star of eight light years and divide it by the six years for the trip recorded by the rocket twin we obtain a velocity one-third greater than that of light! But we are in error here in using eight light years as the distance travelled by the rocket. As a result of the rocket twin's velocity the distance to the star, as measured by him, is fore-shortened because of the Fitzgerald–Lorentz contraction. Thus, using *Equation 1* with the appropriate numerical values which apply here, we obtain a fore-shortened distance of 4·8 light years for the round trip. Dividing this by six years enables the rocket twin to determine his velocity as four-fifths that of light, which is correct.

So far, we have not mentioned the effect which the acceleration and deceleration processes might have on a clock or any time-measuring device or, indeed, on the flow of time itself. We can assume, however, that the acceleration and deceleration processes are small enough here so that the rocket twin and time-recording devices on the rocket are not harmed or altered in any way as a result. The effect of such accelerations and decelerations on the flow of time itself is to introduce an additional time retardation. This is treated under the General Theory, as we shall see in Chapter 5. However, during a long space voyage the time which a rocket spends accelerating and decelerating is very short compared to the total time, so that the total time retardation results almost entirely from the high relative velocity. The results for the particular example cited above, wherein only the time dilation of the Special Theory was considered, are thus essentially correct, and only the Special Theory is needed to discuss the clock paradox fully.

4 · Experimental Proof of the Special Theory

Mass Increase with Velocity

RADIOACTIVE PARTICLES. The first verification of the increase in mass with velocity came from the experimental work of Kaufmann in 1902 and 1906 and, particularly, that of Bucherer in 1909. They were working on something entirely unrelated to relativity – or so they thought. It had been known for some time that certain substances, radium for one, were constantly shooting off three different types of small particles or rays. Such substances are called *radioactive*. They were investigating the particular type of radiation known as *beta rays* and were attempting to determine just what these were. In doing so, they found the velocities with which the individual particles making up the radiation were ejected from different radioactive substances, the amount of electric charge on each, and the mass of each.

The velocities were found to be comparable to the velocity of light; they also found that the *higher the velocity*, the *greater the mass* of the particle. Hence, they obtained many *different* beta particles, each with a different mass. It seemed illogical to these physicists that there could be so many different particles all making up the same beta rays. This also seemed unlikely because these experiments were performed at the time when atomic physics was just being born and most scientists thought all matter was made up of many small particles, most of which were *alike*.

It occurred to these experimenters that the different masses they obtained might be due to the fact that the particles in different substances had different velocities, in which case the Special Theory predicted that their masses *would* be different. So they applied *Equation 2* (see Chapter 3), substituting the mass they determined experimentally for m', the particle's velocity for v, and solved the equation for m, the rest mass of the particle. They found that this rest mass was the *same* for each particle. Furthermore, this mass was the same as that of the electron. And when they found that the charge on each particle was also the same as the electron charge, the conclusion was that the mysterious beta rays were nothing more than electrons shooting out of the radioactive substances at high velocities. This result constituted the first experimental proof of *Equation 2* and the first verification of the Special Theory of relativity.

SOMMERFELD'S THEORY OF ATOMIC ORBITS. The next verification of the mass increase predicted by the Special Theory was in connexion with the improved theory of the atom proposed by Sommerfeld in 1916. Previous to this time, the Bohr theory (1913) had pictured the atom as consisting of a nucleus at the centre with the electrons moving in circles about the nucleus. Sommerfeld, however, showed that it was more correct to assume that, in general, the electron paths were not circles but ellipses, and that the electrons revolved about the nucleus, which was situated at one of the foci of the ellipse, in the same way that planets revolve around the sun, as in *Figure 15(a)*.

It had been shown by Kepler in 1609 that when a planet revolves around the sun, the velocity of the planet changes from a minimum to a maximum and back again during the revolution, the amount of the variation depending on the flatness, or ellipticity, of the orbit. For example, the orbital velocity of the earth varies from about $18\frac{1}{2}$ to 19 miles a second, the variation being small because the earth's orbit is almost circular.

Now, since the velocity changes, the mass-increase formula says that the mass of the electron or planet also should change – and the greater the variation in velocity, the greater will be this change in mass. For the planets, this change is too small to be detected. But the average velocity of the electron in its orbit about the nucleus is about one one-hundredth the velocity of light, so that for a fairly flat orbit the change in velocity, and consequent change in mass, is small but detectable. Sommerfeld showed mathematically that the net effect

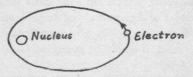

(*a*) Elliptical path for constant electron mass

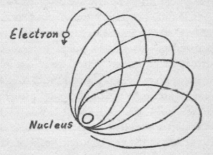

(*b*) Rosette path for variable electron mass

Figure 15. An Electron's Orbit for a Constant and a
Varying Electron Mass

of this change in mass is that the electron will not keep revolving around the nucleus in the same elliptical path over and over again like the earth does around the sun, but that the ellipse will slowly rotate and the electron will describe a rosette pattern, as shown in *Figure 15(b)*. The more technical expression used is that the axis of the ellipse *precesses*.

The verification of the mass-increase effect in the atom thus relied on determining whether the path of the electron about the nucleus was an ellipse, as in (*a*), or a rosette, as in (*b*). An ellipse would indicate that the electron's mass *did not* vary, and the rosette would prove that it *did*, as the Special Theory predicted. It might seem at first glance as if it is impossible to determine the path of a single electron about a nucleus, since not only do we have no way of chopping off a single atom from a substance, but it would be impossible to see such an atom – even with the most powerful microscope we have. Further, since the orbital velocity of the electron about the nucleus is about one one-hundredth of the velocity of light, it would be going too fast for us to see it anyway!

But we do not have to actually see the electron revolving to answer our question as to the shape of its orbit. Fortunately, the particular shape of the orbit produces certain effects which we can examine experimentally. These effects manifest themselves in what is called the *spectrum* of the substance. Most of us know that if we allow a ray of sunlight to pass through a wedge-shaped piece of glass, called a *prism*, the emergent light is found to be broken up into a band of different colours – red, orange, yellow, green, blue, and violet – called a spectrum. The rainbow is an excellent example of a spectrum; here the light is broken up by tiny water droplets in the atmosphere.

In experimental work a prism by itself is insufficient for creating the best possible spectrum, since much greater precision is needed. An instrument called a *spectroscope* is used which contains a prism plus other necessary devices to help gain this high precision. The spectrum produced by a spectroscope shows a number of lines, called *spectral lines*, which are scattered throughout the various colours of the spectrum.

Sommerfeld showed that if the path of the electron about the nucleus is an ellipse, these lines will consist of a number of single lines, as shown in *Figure 16(a)*. Also, for a rosette-shaped orbit these individual lines should really be split, as shown in (b). (The number of individual lines, split lines, spacing between them, etc., will depend on the particular substance. The drawings are illustrative only.)

We see, then, that the verification of the mass-increase effect in the atom depended on the type of lines the spectra produced. If the spectral lines were single, then the path of the electron was an ellipse, and the mass of the electron did not change during its revolution about the nucleus. But if the

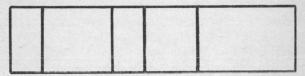

(a) Single spectral lines expected for an elliptical electron orbit

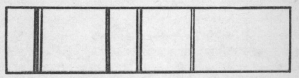

(b) Split spectral lines expected for a rosette-shaped orbit

Figure 16. The Splitting of Spectral Lines in the Mass-Increase Effect

spectral lines were split into two or more, it would mean that the electron path was a rosette, as a result of the variation in the electron's mass, and would verify the mass-increase effect predicted by the Special Theory.

The splitting of the spectral lines was first observed and announced by Paschen in 1916, when he was investigating the spectrum of helium. It is interesting to note that one month after Paschen published his discovery, the Sommerfeld theory was published which predicted the splitting of the spectral lines on the basis of the Special Theory of relativity – and again the mass-increase effect of the Special Theory was verified.

ATOMIC ACCELERATORS. Additional examples of the mass increase with velocity have come from the spectacular results of the giant atom-smashing machines which have been built to investigate the structure of the atomic nucleus. The primary purpose of these machines is to accelerate various atomic particles to high velocities; the more powerful the machine, the higher the velocities; and the higher the velocities, the greater the mass of the particles becomes, according to the mass-increase formula.

Early in 1952 the Brookhaven National Laboratory announced its success in accelerating *protons* (nuclei of hydrogen atoms) up to 177,000 miles a second, which is about 0·95 of the velocity of light. As a result, the mass of the proton was increased to about three times its original mass. And in June 1952 the California Institute of Technology announced it had succeeded in accelerating *electrons* (negatively charged particles with a mass of about 0·0005 that of the proton) to within a tenth of a mile a second of the velocity of light, or approximately $0·9999999c$! The corresponding mass increase was about 900 times the original mass!

The energy of atomic accelerators throughout the world is continually being increased, with the result that the atomic particles used for bullets in nuclear-physics research will have larger and larger effective masses as their velocities come closer and closer to the velocity of light.

One of the theories of light says that light is made up of small particles called *photons* which travel at a velocity of 186,000 miles a second. The question is often asked why photons do not have an infinite mass, since they travel at the velocity of light. The answer is related to the fact that photons are a sort of semi-matter, since they possess only some of the properties of ordinary matter. Their most predominate material property is that of momentum. For example, photons exert a pressure on a surface due to their momentum.

The specific question here is what *Equation 2* predicts for the 'moving' mass of a photon when its rest mass, *m*, is inserted in the numerator. The answer is that the photon simply does not exist at rest, and its rest mass must therefore be considered to be zero. Looking at it another way, electromagnetic waves – the embodiment of photons – only exist in transit. An electromagnetic wave cannot be stationary and still exist any more than a sound wave can. It is true that photons can be stopped – by a light wave striking an opaque surface, say – but their energy is then usually changed into heat at the surface and the photons no longer exist. When a zero rest mass is inserted in the numerator of the mass-increase formula and combined with the zero in the denominator (the result of setting *v* equal to *c*) the mathematical result is an indeterminate value for the moving mass. Thus, the mass-increase formula simply does not apply for photons.

Addition of Velocities

FIZEAU'S EXPERIMENT. We have seen in the previous chapter how the Special Theory predicts that if two rockets, A and B, are travelling towards each other in space, they will each determine their relative velocity by *Equation 4*:

$$V_{AB} = \frac{V_A + V_B}{1 + \dfrac{V_A V_B}{c^2}}$$

We found that this relative velocity is *less* than the sum of A and B's velocities relative to a spacetrian 'at rest'. We have seen, too, that Fresnel's ether-drag theory also predicted that the total velocity would be less than the sum of the two. Fresnel's reasoning had been that objects moving through the stationary ether effectively dragged some of the ether along with them, which made the resultant velocity lower.

If the Fresnel formula is applied to the rockets in space it would predict their relative velocity by *Equation 7*:

$$V_{AB} = V_A + V_B \left(1 - \frac{V_A^2}{c^2}\right)$$

where the expression in parentheses is the Fresnel drag coefficient.

Equation 7 appears to be different from *Equation 4*. Moreover, Fizeau's experiment (see Chapter 2), had shown that Fresnel's formula appeared to be correct. Where, then, does the discrepancy lie? Are we to conclude that *Equation 4*, a result of the Special Theory, is *incorrect* and that *Equation 7*, predicted on the basis of a dragging hypothetical ether, is *correct*? The answer is that *Equation 7* is, in reality, an approximation to *Equation 4*, i.e., starting with *Equation 4*, one can manipulate it mathematically, include a slight approximation, and arrive at *Equation 7*. The relativistic *Equation 4* is thus also correct, and accurately predicts the results of Fizeau's experiment. This experiment, then, constitutes a proof of the relativistic equation for the addition of velocities.

Moreover, the relativistic equation does not postulate the existence of an ether or any reference to any dragging effect. We realize now that the Fresnel drag effect was an artificial device which just happened by accident to agree with Fizeau's experimental results. Fizeau's experiment was repeated in a different form by Airy in 1872 and later by Michelson. Their results also agreed with the equation for the addition of velocities which the Special Theory later predicted.

Equivalence of Mass and Energy

COCKCROFT AND WALTON'S EXPERIMENT. Due to the comparatively large amount of energy which would be released by a relatively small amount of matter, little promise was originally held of ever verifying the mass to energy conversion on a large scale. So scientists concentrated on verifying the equation on an atomic scale – or, more exactly, on a nuclear scale. After the Bohr model of the atom was introduced in 1913, more and more refinements were incorporated until by the late 1920s we had a fairly complete picture of what the atom looked like as a whole. The main attention was then focused on the nucleus itself.

It is now known that the nucleus is made up of what are called *protons* and *neutrons*. Both of these are small particles of about the same mass (about

$$0 \cdot 000000000000000000000001$$

gram each); but the proton has a plus charge while the neutron doesn't have any charge. Furthermore, the lighter the element, the fewer the number of protons and neutrons making up the nucleus, and vice versa. For example, the nucleus of the lightest element, hydrogen, consists of but a single proton, while the uranium nucleus, one of the heaviest, contains 92 protons and 146 neutrons.

The important feature about nuclei that was early recognized was that the protons and neutrons making up the nucleus were held together very tightly. This fact was particularly obvious, since ordinarily we know that two or more plus charges should repel each other – especially when they are very close together as they are in the nucleus. But since they stick together, the nuclear forces holding them there must be much stronger by comparison. Thus, the particles in the nucleus are bound or held together by what is called the *binding energy*. Furthermore, if a whole nucleus could be broken up into smaller pieces (or be made to combine with another nucleus under certain conditions), then this binding energy would escape or be released.

The binding energy which was released was not expected just to appear out of thin air, since one of the unbreakable laws of physics is the *conservation of energy*: *energy can neither be created nor destroyed* but only transformed from one form to another. Where, then, does this energy come from? What supplies it? The answer is furnished by the energy–mass equivalence formula (*Equation 5*) derived from the Special Theory:

$$E = mc^2$$

which says, in effect, that the binding energy released by a nucleus during break-up is supplied by part of the mass of the nucleus.

If a nucleus has a certain mass before break-up, and energy is released during the break-up, then the total mass of the separate pieces must be *less* than the original mass – the lost mass having been converted to energy. If the total mass of the separate pieces were *equal* to the original nuclear mass, then the released energy would appear as if it were suddenly *created out of nothing*, and the law of conservation of energy would be violated. It is also important to realize that in the nuclear processes we know today we can never change *all* the nuclear mass to energy but only the very small part which corresponds to the nuclear binding energy.

To test the above hypothesis, which would verify the energy–mass equivalence, it was necessary to determine accurately the mass of a particular nucleus, then break it apart and determine the binding energy released and the masses of the individual pieces. The first successful experiment which did precisely this was performed by Cockcroft and Walton in England in 1932. They hit the lithium nucleus with a proton. The resulting collision broke the nucleus into two pieces. An appreciable amount of energy was released, and when the total mass of the two pieces was compared with the mass of the original nucleus it was found to be *less*, as predicted. Cockcroft and Walton also measured the energy released during the process, and when this was compared with the computed energy, using *Equation 5*, inserting the value of the nuclear mass which had disappeared, the two were found to agree. Thus, the equivalence of mass and energy was proved twenty-seven years after it was first predicted by Einstein in the Special Theory.

ATOMIC AND HYDROGEN BOMBS. Many experiments have been performed since Cockcroft and Walton's which have further verified the equivalence of mass and energy. These culminated in earth-shaking proofs with the detonation of the first atomic bomb at Alamogordo, New Mexico, on 16 July 1945, and the first full-scale hydrogen bomb at the Marshall Islands in the Pacific on 1 November 1952. Both of these employ the above-mentioned general theory for their functioning. However, there is an essential difference between the fundamental nuclear processes of each.

If you make a graph of the nuclear binding energy released as a function of atomic weight (on the basis of oxygen equal to 16) for the nuclei of all the elements from the lightest, hydrogen (atomic weight 1), to the heaviest, plutonium (atomic weight 239), the result would be as shown in *Figure 17*. (The numbers representing the binding energy released are in arbitrary units for comparison only.) The graph shows that the energy released is *positive* for all elements heavier than silver (atomic weight 108), i.e., if one of these heavier nuclei is broken up or split, energy is released. This release of energy by splitting a nucleus is called *fission*.

Moreover, since the energy released is at the expense of part of the nuclear mass, the total mass of the pieces will always be less than that of the original nucleus. Thus, if we place a single assembled nucleus on one side of a balance scale and the individual pieces after fission has occurred on the other side, we find, in *Figure 18(a)*, that the mass assembled is *greater* than the mass disassembled. The atomic bomb employs the fission process wherein either uranium or plutonium nuclei are split.

91

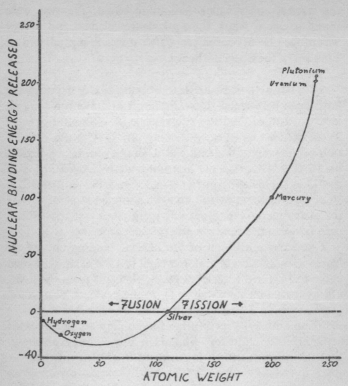

Figure 17. Nuclear Binding Energy Released *vs*. Atomic Weight

For elements of atomic weight lower than that of silver, the binding energy released is *negative*, i.e. energy is not released but *absorbed*. To split such nuclei, an amount of energy numerically equal to this negative energy must be supplied, and no energy is released in the process. How, then, can we get these lighter elements to release energy? By running the process in reverse: instead of having these nuclei absorb energy by splitting them, we can combine two or more of these lighter nuclei into a heavier nucleus

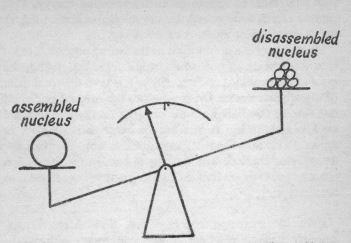

(a) Fission – mass assembled is greater than mass disassembled

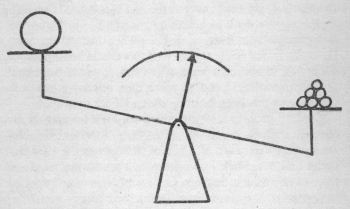

(b) Fusion – mass assembled is less than mass disassembled

Figure 18. Fission and Fusion

with simultaneous release of nuclear binding energy. This process of releasing energy by combining or fusing lighter nuclei into heavier nuclei is called *fusion*.

Since energy is also released in the fusion process, it, too, is at the expense of part of the nuclear masses, so that the mass of the resultant nucleus will be *less* than the total mass of the original nuclei. Or, if we place the original nuclei on one side of our balance scale and the resultant assembled nucleus after fusion occurs on the other side, we find, in *Figure 18(b)*, that the mass assembled is now *less* than the mass dissassembled. The hydrogen bomb uses the fusion process; as the name implies, hydrogen nuclei are fused into heavier nuclei.

ENERGY PRODUCTION IN THE SUN AND STARS. Another striking example of the conversion of mass to energy takes place in our sun. The process by which we get energy from the sun had puzzled scientists for many years. Originally it was thought that the sun was made of coal (or similar substance) which was burning in the ordinary way and giving off its heats as it does when burned on earth. However, it was easily shown that this would be impossible, since, from the known mass of the sun, it would burn itself out in two to three hundred years, and, of course, we know that the sun has been burning much longer than that.

The true process remained a mystery until scientists discovered nuclear processes, with their attendant reaction rates, energy releases, etc. Hence, it was not till 1938 that Bethe and Weizsacker, working independently, produced the true equations for the nuclear processes which applied. In particular, they found that fusion took place – consisting of a chain of nuclear reactions during which four hydrogen nuclei (4 protons) were fused to form a helium nucleus (2 protons plus 2 neutrons). And since the mass of the helium nucleus is about seven-tenths of 1 per cent *less* than the sum of the four hydrogen nuclei, this lost mass is converted to energy. Bethe and Weizsacker computed the rate of energy release for the entire mass of the sun, making use of *Equation 5*, and compared this with the measured amount of radiation we receive from the sun. They found complete agreement between the calculated and measured values, and again the energy–mass equivalence was verified.

Since the energy released by the sun is at the expense of its mass, the sun is slowly but surely eating itself up. At the present rate of hydrogen consumption, the sun uses up about 1 per cent of its mass in a thousand million years; so it will continue to shine for many years. The best 'guestimates', which also consider other factors, conclude that the sun will die out and disappear in from fifteen to thirty thousand million years.

Needless to say, similar fusion processes are responsible for the light emitted by the other stars as well. All the stars will eventually eat themselves up and shrivel down to nothing, their total life span depending on the size of each star originally and the particular processes in each.

The nuclear fusion processes that take place in the stars have been compared with the hydrogen bomb. Although it is true that the process is fusion in both cases, the difference is that the reaction rate for the process is extremely slow in the stars – about a thousand million years – but very rapid in a hydrogen bomb – about a millionth of a second.

THE ATOMIC AGE. Although the first application of the energy–mass equivalence formula in connexion with the atomic bomb was introduced with world-shaking effect, it was but the beginning of what is now being called the Atomic Age. For since that time a great deal of thought and research has been directed towards the peace-time uses of atomic energy, with the result that the increasing number of applications is affecting the scientific progress of many nations in many ways. Most of the applications to date utilize the fission process – but at a much slower rate than in the atomic bomb. Among the most noteworthy of these are the atomic reactors, wherein the energy released by the nuclear reactions is converted into heat. This heat is then coupled to the appropriate devices and used to generate electricity, propel submarines, etc. Of great importance, too, is the formation of radioactive isotopes in the various atomic accelerators throughout the world. These isotopes have extensive uses in medicine, agriculture, and industry in general. In medicine, for example, their radioactivity enables them to be used as tracers in studying the digestive process, blood circulation, etc.

The atomic age has only just begun. No one can envision the many exciting and beneficial additional applications yet to come – all due to the energy–mass equivalence formula derived by the Special Theory of relativity.

Time

IVES'S EXPERIMENT. In the previous chapter we saw that the Special Theory also predicted that if two observers are moving relative to each other, each observer notices that the time processes on the other's system appear to be slowed down – the so-called time-dilation effect.

In practice, this was one of the most difficult predictions to test experimentally, because two observers would have to have a relative velocity comparable to that of light before the effect would be large enough to detect. It was not until 1938 that an experiment was evolved by Ives to test this time-dilation effect.

What was needed first was to get a system to move at a very high velocity relative to the observer. Ives accomplished this by accelerating hydrogen atoms inside a glass tube to a high velocity by means of an electric field. He was able to accelerate the atoms to a velocity of about 1,100 miles a second, or about 0·006 times the velocity of light. Although this was fairly low in comparison with the velocity of light, it was sufficiently high for the looked-for effect to be detected, if it existed.

The problem of determining 'how fast a clock would run' if tied to the moving hydrogen atoms was not as difficult to do physically as it might sound, although high precision was absolutely necessary. The time process that Ives used was the rate at which the electrons vibrated in the hydrogen atoms. We can measure this rate of vibration, or time per vibration, for the hydrogen atoms when they are *not moving*, i.e., at rest relative to the observer, and then again when they *are moving* relative to the observer. And, if the time-dilation formula is correct, the time per vibration in the latter case will be longer than in the former, corresponding to the slowing down of the vibration, and the time-dilation effect would be verified.

To see how these time rates are determined, let us digress for a moment and consider what happens when a tautly stretched wire, such as a piano string, is vibrating. When we strike a key on the piano, the string (or strings) corresponding to that note will vibrate at a particular rate which we refer to as the *pitch* or *frequency* of the note. And, if we strike a note lower down on the keyboard, we hear a note of *lower* frequency, since this string vibrates *more slowly*. Considering the time for a single vibration in both cases, we see that it is longer for the lower-frequency note, since here the string vibrates more slowly. Thus, a *decreased* frequency corresponds to an *increased* or *longer* time per vibration.

The vibrating atom is similar to the vibrating string. A particular string will vibrate at a set frequency depending on its length, tension, etc., which does not change unless the length, tension, etc., change. Similarly, every atom vibrates at its own particular frequency, which should not change.

97

But if it does change, it means that the time processes in the atom have changed. In particular, if the frequency decreases, we have seen this means that the time per vibration has increased, or that time itself has *slowed down* in the particular atom.

Ives measured the frequency of vibration for the hydrogen atoms when at rest and again when they were moving at the velocity of 1,100 miles a second. He found that the frequency decreased, corresponding to an increase in the time per vibration. Furthermore, this increase was exactly the amount given by *Equation 6* for the time-dilation effect when the velocity of 1,100 miles a second of the hydrogen atoms was inserted in the formula. This proved that time did indeed slow down in a system moving with velocity v relative to an observer, and once again the Special Theory was verified.

MESON LIFETIMES. Although it may be a long time before we 'Earthians' can verify the time-dilation prediction by an extended space voyage, a comparable effect has already been observed in a simple, yet very convincing, way. The 'space travellers' here are mu mesons, tiny particles with a mass about 207 times that of the electron. They can be generated in cyclotrons, but mu mesons are unstable. They decay when at rest in the laboratory with a mean lifetime of about 2 microseconds.

Now these same mu mesons are also produced in appreciable numbers by cosmic rays high in the atmosphere. But because of their relatively short lifetime most of them are not expected to arrive intact at the surface of the earth. The puzzling fact is, though, that mu mesons do appear in relative abundance at sea level. In fact, their existence there corresponds to an effective mean lifetime of about 30 microseconds, about 15 times greater than their 'rest' life!

The explanation, of course, is given by the time-dilation

prediction. The high-altitude mesons have an initial velocity of about 0·998 that of light which, when substituted in *Equation 6*, produces a time-dilation factor of about 15. Thus, time on the mesons' 'clock' runs at only about one-fifteenth the rate of earth time and the mesons would be expected to live 15 times longer on the average. Indeed, if an Earthian could be substituted for a meson and travel at a velocity of 0·998 that of light, he, too, would have a mean lifespan about 15 times his present one of about 70 years, or about 1,050 years!

An observer travelling along with the meson would not be aware that time was moving slower on his system unless he compared his clock with one at rest on the earth during his flight. He would attribute the fact that the meson survived the flight from the upper atmosphere to the earth's surface to the apparently short distance of travel. The Fitzgerald–Lorentz contraction foreshortens the travel distance to one-fifteenth the value a stationary observer determines it to be. Thus, a height of 15 miles above the earth's surface becomes only 1 mile for the high-velocity meson, and its survival during this comparatively short distance is not so surprising.

We thus see that one of the most phenomenal predictions of the Special Theory of relativity, the time-dilation effect, is now well established in principle. It only remains to note the effect on a full-scale when 'Earthians' become 'spacetrians' with long space trips a common occurrence.

5 · The General Theory and Experimental Proof

The Principle of Equivalence

SOON after the Special Theory was published in 1905, Einstein turned his attention to phenomena that occurred when observers were not restricted to movement with *constant* relative velocities (i.e., with zero acceleration) but with *varying* velocities (i.e., with accelerations not zero). The results of his reasoning embody the General Theory of relativity, which was presented in 1916. To date, the theory has been verified by three different types of experiments; and, as we shall see, gravitational attraction plays an important part in the theory and its proofs.

Everyone has ridden in an elevator at one time or another and has perhaps noticed that when the elevator is accelerating upwards, he is pushed towards the floor of the elevator. Also, if the passenger is carrying something, it, too, is pushed towards the floor. He feels heavier and everything he carries feels heavier. Furthermore, the faster the elevator starts up (the greater the acceleration), the heavier everything becomes.

Conversely, when the elevator is accelerating downward, everything becomes lighter and the greater the acceleration downward, the lighter everything becomes. In particular, if the elevator should accelerate downward as fast as falling objects accelerate towards the earth (32 feet per second per second), then objects in the elevator would have no weight at all, and everybody and everything would tend to float around in the elevator like soap bubbles! And if the elevator plummeted earthward with an even higher acceleration, everybody and everything would be pressed to the ceiling of the elevator! (Remember, in all cases the elevator is accelerating, i.e., its velocity is changing. When the elevator stops accelerating and moves with constant velocity up or down, these effects will not occur.)

Although the elevator passengers might be flustered momentarily by what is happening, they are not completely in the dark as to the cause of their discomfort. They know that the earth's gravitational force on them has something to do with the peculiar effects they experience. But now suppose these same people are in a rocket travelling out in interstellar space, say on a star-seeing trip. They have no weight, since weight is the force with which a large mass (in our case the earth) pulls on an object, and they are out beyond the pull, or gravitational field, of the earth. Hence, they must be tied down in some way to keep from floating around.

Now, when the rocket accelerates in the forward direction relative to the distant stars, they are pushed back against their seats as in *Figure 19(a)*, and when the rocket decelerates, they are thrust forward in the same way that people are in any conveyance which is speeding up or slowing down on the earth. Thus, the people in the rocket would automatically associate a pressure backward with an acceleration of the rocket, and a thrust forward with a deceleration of the rocket. At other times, when the rocket is neither accelerating nor decelerating, they do not experience any effects.

But now suppose that while they are journeying along in space with a constant velocity relative to the distant stars, a stray planet comes along. It is not seen by anyone in the rocket and narrowly misses hitting the rocket, passing by its tail, as in *Figure 19(b)*. If we assume the rocket motors and controls are such as to keep the rocket moving with constant velocity relative to the distant stars, the question is, what will the passengers feel? With the planet in their vicinity they are again given weight, and they feel this by being pulled towards the planet as it passes, i.e., pushed to the back of their seats. But since they don't know of the presence of the planet and since the effect is the same as if the rocket accelerated, they will erroneously conclude that that is what happened and not give it a second thought.

It might first appear that the passengers and rocket are now in 'free fall' towards the planet so that the people would not seem to be aware of any forces acting on them, but this

★ ★ ★ ★ ★

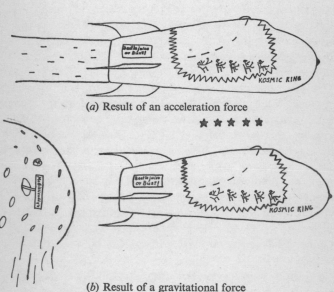

(*a*) Result of an acceleration force

★ ★ ★ ★ ★

(*b*) Result of a gravitational force
Figure 19. The Principle of Equivalence

is not true. We are assuming here that the rocket motors produce just enough force forward to compensate for the gravitational pull backward by the planet on the rocket. This is necessary for the rocket to maintain its constant velocity relative to the distant stars as we have assumed above. (Strictly speaking, *Figure 19(b)* should also show that the rocket motors are exerting a force to combat the gravitational force of the planet even though the rocket does not accelerate in this case since both forces cancel.)

The net effect is that the planet exerts a backward pull on the people, tending to accelerate them towards the planet, while the rocket does not accelerate or decelerate. The effect is the same as if some people were on an elevator on earth travelling upward with a constant velocity and the earth's gravitational field could be turned off and on. While it was off there would be no force pushing the people's feet against the floor of the elevator. As soon as the earth's gravitational field was turned on again the people would be pulled towards the earth and would feel their feet pressed against the floor of the elevator once again.

The larger question is whether there is any way the rocket people can tell (without looking outside) if the forces they feel are due to acceleration or to the gravitational pull of a nearby mass. The answer is that there is no way of distinguishing between the two. Einstein was struck by this equivalence of acceleration and gravitational forces and stated this observation in what is known as the *principle of equivalence – at a single point in space the effects of gravitation and accelerated motion are equivalent* and cannot be distinguished from each other.

Going back to the elevator, we wonder if the apparent increased weight of the passengers, caused by the elevator being accelerated upward, could not also be caused by additional gravitational forces. It certainly could. Suppose that the elevator and its passengers were suddenly 'transplaneted' to Jupiter without their knowing it. They would all feel heavier because Jupiter has a mass over 300 times greater than that of the earth and so exerts a greater gravitational pull on objects on its surface, causing objects to weigh $2\frac{1}{2}$ times more than they do on the earth. Hence, a 200-lb man would weigh 500 lb on Jupiter, and would probably sag to the floor as a result. Furthermore, he would attribute his increased weight to an upward acceleration of the elevator, not knowing that an increased gravitational mass caused it.

Or, if the elevator were transplaneted to Mercury, which has a mass of only $\frac{1}{25}$ that of the earth, everything would weigh only a third as much, and the 200-lb man would weigh only 67 lb. He and the other occupants would interpret their apparent lighter weight as due to the elevator accelerating downward. Again we see that the effects of accelerated motion and gravitation are the same.

It would appear that the principle of equivalence is a simple, or even trivial, observation. It would seem so to one not versed in the history of scientific accomplishment, but Einstein was the first to draw attention to this conclusion. If it had no further ramifications, the principle would be considered interesting and then promptly forgotten. With this principle of equivalence as the basic postulate of the General Theory, Einstein applied a branch of mathematics previously developed by Riemann and others, i.e. tensor calculus, and obtained three important conclusions, each of which was tested experimentally. These will now be discussed in detail.

Newton's Law of Universal Gravitation and Einstein's Theory of Gravitation – Rotation of Mercury's Orbit

Gravitation was a topic which had intrigued people for many years because of the mysterious way in which it acts. A freely falling object always falls towards the earth. But how is it possible for the earth to pull the object towards it without literally reaching up and grabbing the object? The air does not help, because objects are also pulled to the earth even when they are in a vacuum.

Another mystery was the strange force which the sun seemed to exert on the planets to keep them continually revolving about the sun. Kepler had deduced, after considerable observation, that the paths travelled by the planets were ellipses, but he did not know why they were ellipses.

A satisfactory answer to both of these unknowns was given by Newton in 1687 when he published what is now known as Newton's *law of universal gravitation*. This says that every object in the universe attracts every other object with a gravitational force given by *Equation 8*:

$$F = G \frac{mm'}{d^2}$$

where m is the mass of one object, m' that of the other, d the distance between them, and G a universal constant called the *gravitational constant*. When an object is falling freely, the force of attraction of the earth on the object is obtained from *Equation 8* by inserting the mass of the object for m and the mass of the earth for m'. The sun attracts each planet with a force given by the formula, where the mass of the sun is inserted for m and the mass of the planet for m'.

It should be noted that Newton's equation was entirely the result of observation. He observed falling objects and the movements of the planets about the sun and evolved his formula as the one which best fitted the facts – thus obtaining what is called an *empirical* equation. With this equation for the force of attraction between two masses, he then derived the equations for the paths the planets make around the sun. He found that these paths were ellipses, which were *stationary* with respect to the sun, as in *Figure 20(a)*. The planets travelled over and over again in the same elliptical paths in space. Since observations through the years verified this, Newton's law of universal gravitation was hailed as a great accomplishment, as indeed it was.

In his development of the General Theory, Einstein concerned himself with the development of a theory of gravitation. For this reason the General Theory of relativity is also called the *Einstein theory of gravitation.* As a result of this theory, Einstein also determined the equation of the paths the planets make in their journey about the sun. The end result was approximately the same as Newton's but there was a slight difference. Although Einstein also found that the planets' orbits were ellipses, he found that these ellipses were *not stationary* but were *slowly rotating* in space, as in *Figure 20(b).*

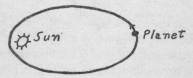

(a) Under Newton's law a planet's orbit is a stationary ellipse

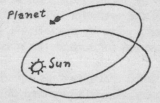

(b) Under Einstein's theory a planet's orbit is a rotating elipse

Figure 20. A Planet's Orbit According to Newton and to Einstein

But this predicted rotation is so slight as to be scarcely detectable for most of the planets. The orbit of the earth, for example, rotates at a rate of only 3·8 seconds of arc every century. When we remember that there are 324,000 seconds of arc in a right angle, it can be realized how small a value 3·8 seconds of arc really is – about one one-hundred thousandth of a right angle. And, further, it take 100 years for the earth's orbit to rotate by this amount. At this rate, it would take about 34 million years for the earth's orbit to rotate one revolution!

Strictly speaking, since the elliptical orbits of the planets

106

are rotating, the orbits are really rosette patterns similar to the paths of the electrons about the nucleus in the Sommerfeld theory. But since this rotation rate is so small, it would take too long for a rosette to be completed. Hence, we think of the planetary orbits as rotating ellipses, and not as rosette patterns.

Obviously, since the Einstein theory of gravitation produces different results from the Newton law of universal gravitation, one of these is incorrect, however slightly. As we shall see, the Einstein theory is the correct one, but we first would like to see how these two theories differ mathematically. If the Newton law of universal gravitation (*Equation 8*) is modified so that the results agree with Einstein's, i.e. so that the orbits of the planets turn out to be slightly rotating ellipses instead of stationary ones, then the corrected formula should be *Equation 9*:

$$F = G \frac{mm'}{d^{2 \cdot 00000016}}$$

which is only slightly different from Newton's law. Thus, Newton's law is correct to a high degree of approximation. It is for this reason that Newton's law of universal gravitation has given such good results for so many years.

One proof of the General Theory of relativity (or Einstein's theory of gravitation) consisted in looking for a planet whose orbit rotated the most over a given period of time. The theory showed that the amount of rotation would be greatest for the planet with the highest orbital velocity. But it was also necessary to use a planet whose orbit was as elliptical as possible, since some of the planetary orbits, like the earth's, are so nearly circular that it is difficult to tell whether or not they have rotated. With an orbit that is very flat, or highly elliptical, it is easy to see in which direction it points, and hence its rotation can be detected.

It so happens that the planet Mercury (the planet nearest

the sun)* has one of the flattest orbits and the greatest orbital velocity. There had been a peculiar and puzzling behaviour of this planet's orbit for many years: it had a rotation of about 43 seconds of arc per century which could not be accounted for. (Although the total rotation of Mercury's orbit is about 574 seconds of arc per century, it was known that 531 seconds of this was due to the gravitational effect of the other planets.) In 1845 the French mathematician Leverrier showed that this excess amount of rotation would be produced if Mercury were being influenced by another planet between it and the sun. The anticipated planet was eagerly looked for by astronomers but has never been found. (The planet Neptune was also predicted by Leverrier as a result of variations in Uranus' orbit, and was thus subsequently discovered. And Pluto, the planet farthest away from the sun, was discovered in 1930 as a result of still-remaining variations in Uranus' orbit.)

The cause of the excess rotation of Mercury's orbit remained a mystery until the introduction of the General Theory of relativity. When the General Theory was used to compute the amount of the excess rotation of Mercury's orbit for the period of a century, the result was 43 seconds of arc—the exact amount of rotation which previously could not be explained. This constituted the first proof of the General Theory. It should be noted that this particular proof is the most convincing of the three types so far known, since the effect is a relatively large one compared to the other two to be discussed.

* Readers may be interested in the following sentence as an aid in remembering the order of the planets: 'Man Very Early Made A Jug Serve Useful, Noble Purposes.' The first letter of each word corresponds to the first letter of the planets in order from the sun outward, i.e. Mercury, Venus, Earth, Mars, Asteroids, Jupiter, Saturn, Uranus, Neptune, and Pluto. If the Asteroids are not to be considered as one of the planets the 'A' may be dropped from the sentence and an 's' added to 'Jug'. (Readers not addicted to the pleasures of the bottle might prefer the following sentence: 'Man Very Early Made A Jug Serve Useless, No-good Purposes.')

We found in the discussion of the Sommerfeld theory of atomic orbits that the rotation of the orbits there was caused by the variation in mass of the electron as predicted by the Special Theory. You may wonder if the excess rotation of Mercury's orbit could be due to the same effect. Since Mercury travels in an elliptical orbit, its orbital velocity varies and, hence, its mass would vary, with its orbit being caused to rotate as a result. However, it can be shown mathematically that the amount of this rotation predicted by the Special Theory will be only one-sixth that predicted by the General Theory. In this case it would account for only about seven of the excess 43 seconds of arc per century that Mercury's orbit is known to rotate. The General Theory, then, satisfactorily accounts for the excess rotation.

Effect of a Gravitational Mass on a Light Beam – Weighing a Beam of Light

In connexion with the General Theory, Einstein also investigated the behaviour of a beam of light under the influence of the gravitational field due to a large mass. His results are best presented by again referring to *Figure 19*, where now we will assume that the rocket passes a row of stars in its flight. Since there is only a single skylight in the ceiling of the rocket, a single beam of light from each star will enter the rocket via the skylight as the rocket passes each star.

In (b) it is seen that while the rocket is passing the row of stars, the stray planet is passing by the rear of the rocket. What effect does this have on the resultant light beam in the rocket? The General Theory states that the gravitational field due to the planet's mass will *attract the light beam towards it* in the same way that the earth will pull a flying bullet or arrow towards it. This causes the light beam inside the rocket to be curved (which is shown highly exaggerated).

It is not surprising to find that a flying bullet or arrow is pulled towards the earth. These have weight (even while in flight), but most people are surprised to find that a light beam also has weight. This did not surprise the scientists of the day, however, because the photons which comprise light beams were considered to have mass, and it was reasoned that if they did, when light falls on a surface these photons would exert a pressure on the surface similar to the pattern of falling raindrops on a roof. This effect has been observed and is known as *radiation pressure*. The pressure is very small, and for the sun's rays on the earth it is less than a tenth of an ounce per acre, but giving a total of almost 300 million tons for the entire surface of the earth. Fortunately, the gravitational attraction of the sun on the earth is many times stronger, so that we are not pushed off into space by radiation pressure of the sun's rays.

In *Figure 19(a)* we see that the rocket is accelerating relative to the distant stars. The effect of this on the light beam inside the rocket is given by the principle of equivalence, which says that the effect is the *same* as it was for the gravitational field caused by the planet's mass. Here, too, the light beam inside the rocket will be curved (again shown highly exaggerated).

To test the prediction of the General Theory that light is deflected by a gravitational field, actually we would have to 'weigh a beam of light'. It isn't possible to catch a lot of photons and pile them on a scale, as we could with bullets, since no one has yet succeeded in building a photon trap. (Moreover, scientists today believe that the rest mass of the photon is zero!) So the photons must be weighed while in flight. This is not difficult to do in theory, since if a beam of light is affected by a gravitational field, its path will be curved, and this is easy to determine, providing the curvature is sufficient. But if a beam of light is not affected by a gravitational field, its path through it will be a straight line, which is also easy to determine.

110

Since all objects on the earth fall about 16 feet during the first second of fall (neglecting the effects of air friction), it is expected that a light beam travelling parallel to the earth's surface would also fall, or bend, towards the earth this amount during the first second of travel. But a light beam also travels 186,000 miles during this time, so it is well nigh impossible to detect such an effect on the earth. However, in our solar system there is a mass whose gravitational pull is much greater than that of the earth, and so the bending would be much greater. This mass is the sun's, which is over 330,000 times greater than that of the earth. Its average density is about a fourth that of the earth, with the overall result that the gravitational pull at the sun's surface is about twenty-seven times that on the earth's surface. Since this is also over ten times that of Jupiter, the largest planet in our solar system, the bending of a light ray due to the sun's gravitational attraction will be greater than for any other object in our solar system. Hence, the sun is the best 'scale' to use for the weighing.

The beam of light must, of course, come from a star. The procedure for weighing is illustrated by *Figure 21*. The initial position of the star A, whose light is to be weighed, is shown in (*a*). There are no intervening gravitational masses, so the star's light travels in a straight line from the star to the observer on earth. Later on, the earth has travelled sufficiently far in its orbit so that the sun comes between the earth and the star in such a position that the light from the star just grazes the sun's surface on its way to the observer on the earth, as in (*b*).

But here a difficulty is encountered, for if the star's light is just grazing the sun's surface the observer will not be able to see the star, since the sun's light will be too bright. The only solution is to observe the star's light grazing the sun during a total eclipse of the sun when the moon blots out the sun's light completely, as illustrated. For this reason

111

Einstein suggested that this effect be looked for during a total solar eclipse.

Since the deflection of the star's light while grazing the sun is so slight, precise photographic techniques are necessary. The actual procedure consists of photographing the star while in position (*a*), showing its position relative to its neighbouring stars, and then again at a later time during a total solar eclipse, as in (*b*). In this latter position, it will ap-

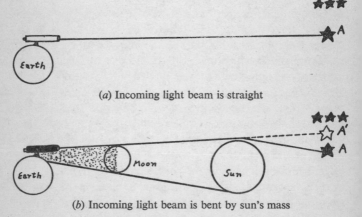

(*a*) Incoming light beam is straight

(*b*) Incoming light beam is bent by sun's mass

Figure 21. Weighing a Beam of Light

pear to the observer (and his camera) that the star is at the apparent position A'. A picture taken at this position is compared with that of position (*a*). The comparison should show that the star appears to have moved, provided that the sun's mass has deflected the star's light as predicted by the General Theory. In particular, Einstein predicted that the deflection would be 1·74 seconds of arc for a light beam grazing the sun's surface.

The most favourable total eclipse of the sun after the General Theory was presented in 1916 occurred on 29 May 1919. This eclipse was particularly favourable for the test because the earth and sun happen to be lined up with a patch of bright stars at the end of May every year, and so there were a number of stars to choose from during this eclipse. Accordingly, two British astronomical expeditions were fitted out. One, under A. C. Crommelin, went to Sobral in northern Brazil, while the other, under A. S. Eddington, went to the West African Isle of Principe in the Gulf of Guinea. Both groups photographed a number of stars, and upon their return to England the photographic plates were developed and compared with pictures taken when the sun was *not* in the same stars' vicinity.

The Sobral group found that their stars had moved an average of 1·98 seconds of arc, and the Principe group that theirs had moved 1·6 seconds of arc. This nearness to the 1·74 seconds of arc predicted by Einstein was sufficient to verify the effect. Since then, more than ten different results have been reported which also confirm the prediction.

It is interesting to speculate how massive a star would have to be so that its gravitational attraction is strong enough to prevent *any* of the star's light from leaving the star. It can be shown that for a star of the same radius as the sun this would occur if its mass were approximately 400,000 times the sun's mass. If such stars existed, we would never be able to see them, regardless of how close they were or how brilliantly they shone!

Using the Newton law of universal gravitation, a value can also be obtained for the bending of a light ray. This turns out to be exactly one half the value given by the General Theory of relativity; and so for the sun the value would be 0·87 seconds of arc. None of the experimental determinations to date has been of this order of magnitude – all being larger and within a reasonable range of the Einstein value. This again points up the slight difference between the Newton law and the Einstein theory.

Effect of a Gravitational Mass on Time – Slowing Down of Atomic Clocks on the Sun, Stars and Earth

Another result of the General Theory is the effect of a gravitational mass on time. The prediction is that all time processes will be slower on a large mass than on a small mass, or that time will move more slowly on a relatively larger planet, such as Jupiter, than on earth. Although a clock which runs at a certain rate on the earth will run slower on Jupiter, it will run even slower on the sun. Indeed, Einstein found that a second of time on the sun should correspond to 1·000002 earth seconds.

To measure this slight difference, literally speaking, we would have to put a clock on the sun, synchronize it with one just like it on the earth, and then periodically compare the two. With the difference in time rates indicated, the sun clock would be one second behind the earth clock after 500,000 seconds, or after just under six days. Of course, we have no way of putting a clock on the sun; but we do not have to, since we have many 'atomic clocks' there already. These are the vibrating atoms discussed in the previous chapter in connexion with Ives's experiment.

This particular prediction of the General Theory can be tested by the same method Ives used in searching for the time-dilation effect predicted by the Special Theory. Since the light from the sun is caused by many different types of vibrating atoms, the frequencies of these vibrations can be determined experimentally, from which the times per vibration can be computed. The frequencies and corresponding times per vibration also can be measured for the same atoms vibrating on earth. These can then be compared with the former. As we saw previously, if the frequencies of vibration of the atoms in the sun are *less* than those for the same atoms on earth, it means that the times per vibration have *increased*, or that time itself is *slowed down* on the sun.

Since the prediction was that the frequencies of the sun's light would be decreased, the frequencies were expected to be shifted towards the red end of the visible spectrum, because the frequency of the colour red is lower than that of the other colours in the spectrum. To differentiate this particular red shift from other effects which also produce a red shift, it is referred to as the *relativistic* or *Einstein shift*.

The Einstein shift was first looked for in the sun. Unfortunately, the expected shift was so small that it was barely within the limits of measurement, so that these early attempts did not confirm the effect conclusively.

Since then, a class of stars called the *white dwarfs* has been used to detect the shift. These white dwarfs are small in comparison with most stars, but extremely dense. In particular, B Sirius (the companion of Sirius, the Dog Star, which is really a double star) has a diameter about 3 per cent that of the sun's, but its density is more than 25,000 times greater. On such a star a pint of the nuclear fluid making up the star would weigh about 18 tons! You can expect that life *would* be slower on such a star, where a person would be so crushed by his own weight that he couldn't even move!

Since the predicted frequency shift for the star B Sirius is over thirty times that expected for the sun, this star was used by Adams in 1925 in his attempt to find the predicted effect. He found a frequency shift towards the red end of the spectrum of the expected amount. Then, in 1962, J. E. Blamont and F. Roddier of the Meudon Observatory in France reported the result of a very accurate measurement of the Einstein shift in the sun. The amount of the shift was almost exactly the amount predicted. These experiments constitute proof that a strong gravitational field *does* slow down time processes, as the General Theory predicted.

But the relativistic red shift should result from any mass, theoretically, regardless of how small. Thus, the mass of the earth is sufficient to product the effect, but the amount would be considerably smaller even than that resulting from the sun, whose mass is about 300,000 times that of the earth. It would be ideal, too, for the effect to be detected at or near the earth's surface so an experiment could be performed in an earth-based laboratory. Otherwise, satellites would be necessary, and such an experiment would be very expensive and complicated.

The fractional change in gravitational energy near the earth's surface is only about one part in about ten thousand billion, and so when the Einstein shift was first predicted it was deemed too small to be measurable on the earth's surface. But in 1958 the German physicist Rudolf L. Mossbauer discovered what is now called the *Mossbauer effect*, which has made possible the detection of the effects of such extremely small energy changes. In fact, the effect has also made possible such a large number of highly precise experiments of such hitherto undreamed-of sensitivity that Mossbauer was awarded the Nobel prize in physics in 1961 for his discovery. The gamma rays, or photons, emitted by identical nuclei all have the same energy levels, characteristic of that particular nucleus. Conversely, such nuclei will absorb photons of the same amounts of energy and only these precise amounts. In other words, the frequencies of the gamma rays emitted are the same for all the nuclei because the frequency is directly proportional to the energy. Thus, two identical nuclei can be arranged to have one act as an emitter and the other as an absorber. When a photon emitted by one is captured by the other they are said to be in *resonance*. In this respect the two nuclei act like two piano strings which are tuned together or even two clocks which are ticking in unison.

In actual practice it is difficult to place two (or more) identical nuclei in resonance because the emitting nucleus experiences a recoil when it throws out its photon just as a rifle does when it fires a bullet. This recoil imparts a backward velocity to the nucleus, with the necessary energy being supplied by the emitted photon. Thus, the emitted photon's energy is decreased, and since it is then insufficient to permit capture by the absorbing nucleus the resonance is destroyed. Or, considering the result in terms of the frequencies of the gamma rays, again similar to the frequencies of two vibrating piano strings, the resonance is 'detuned'.

Mossbauer supplied the clever innovation of attaching the emitting nuclei to solid material which effectively increases the mass of the emitting nuclei to the point where the recoil energy is practically nil. Thus, the emitted photon's energy is not diminished, the frequency of the gamma ray is not changed, and resonance is maintained. The advantage here is that if the photon's energy is sufficiently 'sharp', i.e. if it is always approximately the same precise amount for identical nuclei, then the emitter and absorber nuclei are very finely tuned. Then only a very small change in the energy of the emitter or absorber nucleus, or both, will 'detune' the nuclei and destroy the resonance.

It should then be obvious why the Mossbauer effect comprises such an effective technique in the detection of extremely small changes in energy or frequency. In principle, one has two nuclei 'tuned' to each other, and then if the energy of one (or both) changes because of some effect or phenomenon being sought, the phenomenon manifests itself by the resultant 'detuning' of the nuclei which occurs.

The change in gravitational energy near the earth's surface and the Einstein shift produced by it is easily within range of techniques employing the Mossbauer effect. In fact, the Mossbauer effect can be used to detect the velocity of an object as low as an inch per 1,000 years! The effect is thus ideal to detect the gravitational red shift predicted by the General Theory, and such a test was proposed soon after the discovery of the effect.

The first successful experiment was performed at Harvard University in 1960 by Dr Robert V. Pound and his graduate assistant Glen A. Rebka, Jr. They used a tower 74 feet high with radioactive cobalt-57 decaying by gamma rays, or photon emission, to iron-57 as the emitter and absorber nuclei. This particular nucleus was selected because one of its sharply defined energy levels makes possible very fine tuning.

The emitter nuclei which ejected photons were placed at the bottom of the tower with the absorbing nuclei at the top. The absorbing nuclei did not capture the photons travelling up the tower, i.e. no resonance was observed, thus indicating that the resonance was 'detuned'. But when the absorbing nuclei at the top were given a velocity downward towards the source nuclei the resonance was restored.

The reason is that the energy of the photons was decreased because of the action of the earth's gravitational field on them. The photons thus had their energy and corresponding frequency shifted, which caused them to be 'out of tune' with the absorbing nuclei. But when the energy of the latter was shifted the correct amount by their motion they were then 'in tune' again with the approaching photons and resonance was restored.

Since the frequency of photons, or gamma rays, is proportional to their energy, the decreased energy produced a decreased frequency. And since wavelength is inversely proportional to frequency, the decreased frequency resulted in an increased wavelength. Thus the wavelength was shifted to the red end of the spectrum. Further, the very small amount of the red shift detected was the same as that predicted by the General Theory (within the limits of error of the experiment), and the prediction was verified by the Mossbauer effect in a most conclusive way.

Now, the ejection of photons can be viewed as nuclei which are 'vibrating' with the emission of a single photon corresponding to one vibration. Then the frequency of the emitted photons, or gamma rays, refers to the rate of vibration. In this sense, too, the vibrations of a nucleus are like the ticking of a clock, and when the frequency of vibration decreases the nuclear 'clock' is slowing down. Thus, a decreased frequency is interpreted as a slowing down of time itself. Pound and Rebka's experiment, therefore, verified the predicted retardation of time as a result of a gravitational field, in this case the earth's.

Pound and Rebka also interchanged the positions of the emitter and absorber nuclei and measured the shift in the wavelength of the light going *towards* the earth down the tower length of 74 feet. As expected, they found the opposite effect, i.e. the wavelength of the light was shifted towards the blue end of the spectrum by the amount predicted by the theory. This result made the results of the entire experiment even more conclusive, and the Einstein shift predicted by the General Theory is now believed to be verified conclusively.

6 · Relativity and the Nature of the Universe

Types of Universes

THE nature of our universe is a fascinating topic which has stirred the imagination of people for many years. Much speculation has produced many different possible universes as prototypes of our own, but in this chapter we shall confine ourselves to the particular type of universe which the General Theory of relativity originally suggested ours is. Strictly speaking, this topic is called *relativistic cosmology*. However, before we can discuss the universe we live in, we must examine the various types of universes that can possibly exist. For simplicity, we will first look at possible worlds which can exist in one dimension, then two dimensions, and so on.

As an example of a *one*-dimensional world, consider *Figure 22(a)*. First assume we have 'a one-dimensional' bug living in a one-dimensional world, which is nothing more than a piece of a straight line on which he is confined. He cannot move sideways or up and down but only backward and forward. Since he is confined to a piece of line which has a definite, measurable length, his world is said to be *finite*. And since he can't keep going in one direction (he is stopped by the ends where he must back up), his world also is said to be *bounded*. Thus, a bug on a line segment lives in a one-dimensional world which is *finite and bounded*.

If we now put our bug on the perimeter of a circle, he can still move only forward and backward. However, he can move this way indefinitely, without ever being stopped by any barrier. His world is now *unbounded*. But since the length of the circle is a definite, measurable length, his world is still finite. Hence, a bug confined to the perimeter of a circle

Finite and bounded

Finite and unbounded

(a) One-dimensional worlds

Finite and bounded

Finite and unbounded

(b) Two-dimensional worlds

Finite and bounded

Finite and unbounded

(c) Three-dimensional worlds

Figure 22. Various Types of Universes

lives in a one-dimensional world which is _finite and unbounded_.

We could put the bug on a straight line which is infinitely long (or on a circle of infinite radius), in which case his one-dimensional world would then be _infinite and unbounded_.

Figure 22(b) shows our bug in several _two_-dimensional worlds. When he is confined to the surface of a square, he can move in any direction backward or forward and also sideways. But he cannot move off the surface. Since the area of the square is measurable, his world is finite; and since he can't keep going in a straight line through the edges of the square, his world is bounded. Here, then, his two-dimensional world is _finite and bounded_.

If we put the bug on the surface of a sphere and don't permit him to leave the surface, you can see that the bug's two-dimensional world is _finite and unbounded_. Furthermore, if we put him on an infinitely large flat plane, his two-dimensional world will then be _infinite and unbounded_.

We should be grateful that we do not live in a two-dimensional world. Life would be very 'plane' indeed, since everything about people would always be flat, including their heads, their pocket-books, and their beer! (The latter would have to be purchased in sheets in the same way we now purchase sheets of stamps from the post office.) And even worse – people would be only living shadows of themselves! These are the 'plane' facts.

It becomes increasingly difficult to visualize examples of various types of worlds of increasing number of dimensions, and you will have to help the illustrations along with your imagination. The three-dimensional examples about to be discussed are _not_ representative of our universe, since we have seen that our universe should be thought of as containing four dimensions (time being the fourth).

Examples of two types of _three_-dimensional worlds are shown in _Figure 22(c)_. It is assumed now that the bug is in empty space, with himself as the only occupant. If we put him inside a hollow, spherical shell, his three-dimensional world will be *finite and bounded*. It is finite because the volume of the sphere is finite, and bounded because he cannot keep going in a straight line indefinitely but will be barred by the inside wall of the spherical shell. But now, being in a three-dimensional world, he can move up and down in addition to moving forward, backward, and sideways.

To construct a hypothetical three-dimensional world which is *finite and unbounded*, we will assume that our bug lives with a whole family of bugs in a space which has no physical boundaries or barriers. If we further assume that the bugs are very massive, then none of the bugs will be able to leave the group because the gravitational attraction of the group as a whole on each bug will prevent it. Furthermore, since the gravitational attraction is so strong, light rays will not be able to leave the mass of bugs either. Thus, even if a bug looks off in the direction of space beyond the group, his line of sight will curve back towards the group, always producing 'bugs in his eyes', and he will never be able to see beyond the group. 'Straight ahead' for each bug always will mean towards the centre of the group. The bugs will not be conscious of any physical barrier, though; as far as they know, they will live in a world which is unbounded. Their world is finite, since the size of the group as a whole is finite and the group constitutes their world.

An example of a three-dimensional world which is *infinite and unbounded* could exist for a bug if we left him alone to roam all by himself in an infinite space without any gravitational masses or other forces to hinder him. Or, if there were other bugs present, their universe could still be infinite and unbounded in an infinite free space, provided that gravitational attraction could be turned off and on like other types of physical attraction.

The General Theory and Our Universe

Newton originally considered the universe as consisting of a finite amount of matter spread uniformly throughout a finite amount of space. He quickly rejected the possibility, however, because the law of universal gravitation predicted that the mutual gravitational attraction of every particle to every other particle would cause all of the matter of such a universe to be drawn together. The result would be a single very large sphere containing all the matter of the universe located at rest at the centre of the finite space. But the universe is not like this at all, as Newton knew.

So Newton revised his theory and assumed the universe to be infinite in extent. Scattered throughout the vast void at very great distances from each other were an infinite number of large gravitational masses – the stars, galaxies, and clusters of galaxies.

But on the basis of the General Theory of relativity, Einstein was able to show that such a universe as depicted by Newton was unlikely, if not impossible, for mathematical reasons. In particular, he showed that in such a universe the average density of matter throughout the entire universe would have to be zero. Thus, according to Einstein, Newton's universe 'ought to be a finite island in the infinite ocean of space'.

The laws of Newton were predicated on the fact that light travelled in a straight line. The General Theory showed, however, that light rays are deflected by gravitational masses. On the basis of the results of the General Theory and his reasoning, Einstein originally concluded that our universe is *finite and unbounded*.

Our universe, according to Einstein's original theory, is analogous to the surface of a sphere in two dimensions, which is finite and unbounded. If we travel in a straight line on the surface of a sphere (the earth, say), we will eventually come back to the starting-point without having consciously turned around anywhere during the journey. A straight line on the earth is one that follows the earth's surface. We know that the earth's surface is round, but we cannot detect this easily with the eye, since the curvature is so slight.

Out in space a straight line is determined by the path a light beam takes. When it travels far away from any gravitational masses it is not influenced by them; but in their vicinity the light is curved or bent towards the masses. For this reason, space itself is said to be 'curved'; hence the origin of such terms as *space-curvature*, or the *curvature of space*. Space should not be imagined to be curved in the ordinary sense of the word, but only in that it contains gravitational masses (stars and other solar systems that may exist) which cause light rays to be deflected in their vicinity.

The property of gravitational masses to deflect light rays explains why our universe is unbounded. For, although light rays travel in straight lines in the vast reaches of space between the stars, they will be deflected when passing near the stars. And, if light rays suffer enough successive deflections, they can be caused to turn completely around and face in the opposite direction, in the same way the traveller does when he is half-way around the earth. And, like the earth voyager who returns to his starting-point by continually travelling in a straight line on the earth's surface, a space traveller in our universe would also find himself back at the earth if he travels what appears to him to be a straight line in space. He would no more know that he is travelling a gigantic circle in space than the earth voyager is conscious of travelling in a circle on the earth. In general, a straight line in space, then, is the path a light beam takes, which may be straight, or curved, or a combination of both. In order to avoid confusion with what we ordinarily think of as a straight line, we will refer to the lines in which light travels as *space lines* rather than straight lines in space.

So if the Einstein conception of the universe is correct, a

'spacetrian' who leaves the earth and continually travels in a space line will *always* end up again at the earth, regardless of what his original direction is away from the earth. And again like the earth voyager, no barrier of any kind will be met during such a trip around our universe; hence our universe is unbounded.

Our universe is finite because if you continually travel in a space line and end up at the starting-point again a certain time later, only a finite amount of space would be passed through. And again like the surface of the sphere, this amount of space should be measurable.

The physical picture of our universe is that of a vast ocean of space with galaxies of stars (plus whatever other celestial material there might be) embedded in it in a more or less uniform distribution, like raisins in a loaf of raisin bread. (Some refer to the gravitational masses as 'pimples in space', 'kinks in space', or 'ripples on the space surface'.) Moreover, there is no outside edge to the universe, for we have seen that continual travel in a space line brings you back to the starting-point. Our universe *closes on itself*.

A crude representation that might help to visualize this original Einstein conception of our universe is shown in *Figure 23*. The earth is shown at the centre. (It must not be concluded from this that the earth is actually at the centre of the universe, since there is no such thing as a centre of our universe any more than a centre exists on a two-dimensional spherical surface.) If you travel outward from the central earth on one of the space lines (represented by the radial lines extending outward from the central earth) you will keep getting farther and farther away from the earth. On the two-dimensional spherical surface this corresponds to getting farther and farther away from your starting-point as you travel in a straight line on the earth's surface. At a certain distance, represented by the large circle, the spacetrian will be at the maximum distance from the earth, in the same way that the earth traveller is at a maximum distance from his starting-point when he has reached a point on the earth diametrically opposite his starting-point.

As the spacetrian keeps on travelling along his space line, he will now find himself approaching the earth again. This

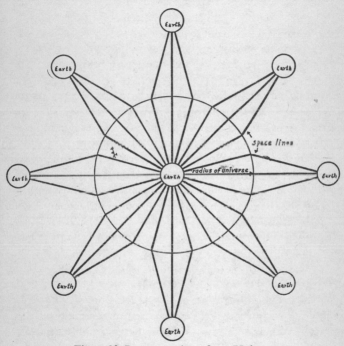

Figure 23. Representation of our Universe

is shown by the portions of the space lines outside the circle converging towards the other earths. These earths really represent our earth at the centre of the diagram, but to the space traveller who doesn't understand the nature of our universe they will look like duplicates of our earth until he 'gets down to earth' and finds that it is really the same earth he took off from. In the two-dimensional analogy, the earth traveller, too, eventually sees his starting-point in front of him when he knows he left it behind him. If he did not understand the nature of his world he, too, would think he was seeing a duplicate of his starting-place.

Since there exists a maximum possible distance from the earth, we can look upon this distance as being the *radius of the universe*. This is again analogous to the spherical surface: for every point on it there is another point which is a maximum distance from it located diametrically opposite. The distance between the two points will, of course, depend on the radius of the sphere – the larger the radius, the greater the distance between the two points. For a sphere the size of the earth, for example, the point which is at the maximum distance from the North Pole is the South Pole.

In *Figure 23* the radius of the universe is the radius of the big circle. Einstein was able to derive an expression for the radius of the universe on this basis; he found that it depended on the average density of matter in the universe. (Mathematically speaking, he found that the radius varies inversely as the square root of the density.) Using the best 'guestimate' for the average density of matter in space, the present conclusion is that the radius of the universe is about 200,000,000,000,000,000,000,000 miles.

We can conclude that according to the General Theory of relativity, the universe was considered to be finite and unbounded. Whether it is or not may never actually be determined experimentally. However, it is amusing to predict what could take place many years from now if it actually were. An astronomer might some day build a super-duper telescope, and we can imagine what will happen when he looks through it. He may see a shiny luminous object which looks like the moon, but with a very peculiar-looking curved tree growing out of it. Only after many hours of quiet and careful scrutiny will it dawn on him that he is looking at his

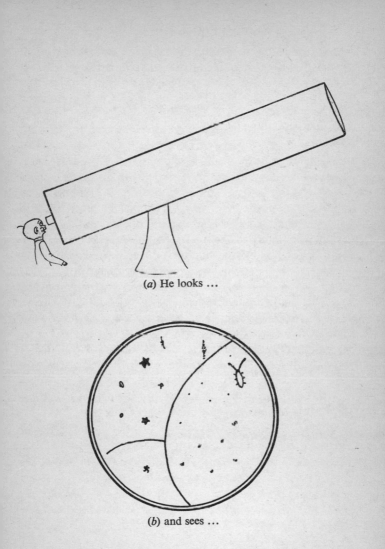

(a) He looks ...

(b) and sees ...

Figure 24. If the Universe is Finite and Unbounded ...

own gleaming bald pate, the light from which has gone completely around the universe and returned! (See *Figure 24*.)

The Expanding Universe

Although the Einstein model of the universe was an intriguing one and was based on a rather firm mathematical foundation, an important development occurred in 1929 which completely invalidated it. For it was during that year that the American astronomer, Edwin P. Hubble, announced that on the basis of experimental evidence (the so-called 'red shifts') it appeared as if all the other galaxies were rapidly running away from us. The interpretation of this is that our universe is in a state of very rapid expansion.

This development nullifies the original Einstein model of the universe as elucidated in the preceding section because it was based on our universe being static, i.e., not expanding. There have been a number of dynamical models of the universe put forth in recent years which incorporate the expanding-universe feature, several of which are as tantalizing as the original Einstein model. The interested reader can read about these in the many excellent books currently available on astronomy and cosmology, since further pursual of this topic is beyond the scope of this book.*

* See, for example, the present author's *Early Theories of the Universe* (1967) and *Modern Theories of the Universe* (1963).

7 · The Unified-Field Theory

Now that the discussion of the theory of relativity has been completed, it may look as if the story has been told and there is nothing more that can be said. Nothing could be farther from the truth. The theory of relativity was only an introduction to the much bigger and even more tantalizing problem with which Einstein grappled for the last twenty-five years of his life. This is called the *unified-field theory*.

The problem is easy to understand but difficult to solve. We will remember that one of the basic phenomena in our universe is that of gravitational attraction, i.e., every object in the universe attracts every other object. We saw that this could be expressed mathematically by Newton's approximate law (*Equation 8*):

$$F = \frac{G\, mm'}{d^2}$$

where m is the mass of one object, m' that of the other, d the distance between them, and G the gravitational constant.

But we also know of other types of forces which are similar to gravitational attraction. Two unlike electric charges (a negative charge and a positive charge) will also attract each other with a force given by *Equation 10*:

$$F = \frac{C\, qq'}{d^2}$$

where q is the amount of the negative charge, q' that of the positive charge, d the distance between them, and C a constant. This particular formula is called *Coulomb's law*, after its discoverer. And we also have a similar formula giving the force of attraction between two unlike magnetic poles (a north magnetic pole and a south magnetic pole), *Equation 11*:

$$F = \frac{K\,MM'}{d^2}$$

where, similar to the other two formulas, M is the pole strength of the north magnetic pole, M' the pole strength of the south magnetic pole, d the distance between them, and K another constant, different from the previous G and C.

Comparing these three equations, there are two important conclusions. First, the three equations which mathematically express three *different* and entirely unrelated phenomena are *identical in form*. The second important conclusion to be drawn is the difference between the gravitational force of two masses, on the one hand, and the electric and magnetic forces, on the other. Gravitational forces are forces of *attraction only*, but electric and magnetic forces can be of *attraction or repulsion*. For example, two unlike electric charges will attract each other, but two like charges (two negative charges or two positive charges) will repel each other. Similarly, two unlike magnetic poles will attract each other, but two like poles (two north poles or two south poles) will repel each other. The three types of forces are similar in one respect but dissimilar in another.

Historically, these equations were evolved empirically by different people working entirely independent of one another. But the similarity between the three types of force equations (gravitational, electric, and magnetic) is so striking that it seems as if all three must be branches of a more fundamental or basic phenomenon of nature. The attempt to derive these equations from more fundamental theory comprises one particular aspect of the unified-field theory.

The general purpose of the unified-field theory is a much broader one than this, however. For it is an attempt to deduce *all* the physical phenomena we know of from a few simple fundamental principles. Up to the present time, the laws of physics have been developed in separate sections or branches in a generally unrelated way. The laws of thermodynamics comprise one branch, those of optics another, etc. As we have matured scientifically through the years and our store of knowledge of the physical world has increased from an infinitesimally small amount to an infinitesimally larger amount, we have seen interrelations between the branches. These amalgamations, when they occur, have enabled us to gain scientific knowledge at a much faster rate. If, through the unified-field theory, the fundamental laws of the universe can be stated for all time, then the laws of all the various branches should flow as an effortless consequence.

Although a unified-field theory sounds highly desirable, how do we develop such a theory in actual practice? As the name implies, the theory concerns itself with *fields*. When two gravitational masses (or electric charges or magnetic poles) attract each other, the interaction takes place in the region, or field, between the masses. Since objects are influenced by other objects at a distance from them, Newton called such forces 'action at a distance'. What happened in the region between the objects was not known. Einstein's approach to the problem was to consider the field itself in an effort to understand the basic underlying properties of fields in general. Then gravitational, electric, and magnetic fields would follow as special cases and the General Theory of relativity (since it is a theory of gravitation) would be derivable from the unified-field theory.

In 1953, two years before his death, Einstein announced the results up to that time of his search for the ideal field theory. He believed he had succeeded in uniting the phenomena of gravitation and electromagnetism into a single theory. Unfortunately, the set of equations his theory generates has an infinite number of solutions, and there is no way of determining which solution is correct and applies to our universe. Or no way has yet been found to verify the theory experimentally. So no one knows whether the Einstein unified-field theory, in its present form, is correct or not.

Up to now, scientists have been concerned mainly with directly measurable quantities, such as temperature, pressure, force, etc., and have evolved theories in terms of them so that they can measure these things experimentally in an almost mechanical aftermath. The emphasis has not been on the understanding of a phenomenon but on the physical proof of it, or as Einstein called it, the 'closeness to experience'. Admittedly, physical proof is desirable, but it should be emphasized that it is not necessarily the most important element. In all probability, the ultimate unified-field theory will not be as amenable to experimental proof as other physical theories have been, because of the more subtle nature of the phenomena with which it deals.

But the tremendous power of a successful unified-field theory can be seen in that it produces much more fruitful results than does the mere repetitious derivation of formulas we already know. For if the underlying theory of fields is understood completely, we should be able then to understand other forces which also exist at the present time but about which we know very little.

An example is the powerful forces that hold nuclear particles together, i.e. *nuclear forces*. We know that these forces are much stronger than the coulomb forces, whereby like charges tend to repel each other – but beyond that we know little. It may very well be that there exists such a thing as *nuclear fields*, analogous to gravitational fields (but much stronger), which the unified-field theory can predict.

With a slight stretch of the imagination we can envisage an even broader usefulness for the unified-field theory. We have seen that gravitational forces are forces of attraction only, while magnetic and electric forces are forces of attraction and repulsion. Is it not possible that a gravitational force of *repulsion* can exist in our universe and that this missing link only awaits our discovery of the general laws of fields before we can create such a repulsive gravitational force?

Such reasoning can be extended further. Perhaps a true understanding of fields will enable us *to predict and create* fields which are completely different from any that we now know. Such scientific progress as this is rare in our civilization, but a brilliant example occurred when Maxwell in 1864 mathematically predicted the existence of radio waves from the most elementary knowledge of electricity and magnetism. The fact that these could not be produced experimentally for another twenty-odd years makes his achievement all the more notable. We may be on the verge of similar developments today through the unified-field theory.

We see that we are living in exciting times. Not only is scientific progress being made at an ever-increasing rate, but every new scientific discovery predicts many more to come – each more enthralling than the last, and for many of these we will be indebted to the genius of Albert Einstein and to his theory of relativity.

Figure 25. The 'Spacetrian' of the Future

Index

Index

139